To Bob Drenten with
warmest regards,

Stuart W. Scott

Charles J Cooper

# Exploring
# The Unknown
## Great Mysteries Reexamined

# Exploring
# The Unknown
## Great Mysteries Reexamined

**Charles J. Cazeau**

and

**Stuart D. Scott, Jr.**

*State University of New York at Buffalo*
*Amherst, New York*

**PLENUM PRESS** • **NEW YORK AND LONDON**

Library of Congress Cataloging in Publication Data

Cazeau, Charles J
  Exploring the unknown.

  Bibliography: p.
  Includes index.
  1. Civilization, Ancient. 2. Curiosities and wonders. 3. Flying saucers. I. Scott, Stuart
D., joint author. II. Title.
CB311.C37                        001.9'42                         78-27413
ISBN 0-306-40210-6

© 1979 Plenum Press, New York
A Division of Plenum Publishing Corporation
227 West 17th Street, New York, N.Y. 10011

Printed in the United States of America

This book is dedicated to
Stuart D. Scott, M.D., Floyd A. Cazeau, and Howard G. Carroll,
who would have found themselves among these pages.

# Preface

The purpose of this book is to explore some of those great mysteries of the earth that have captured the popular imagination, and especially those having their roots in our specialties of archaeology and geology. The average reader probably is unfamiliar with the earth sciences or the archaeological history of man. Nor does the average reader have the time and literary resources to verify all he or she reads. Our aim is to lend a helping hand by examining the evidence that surrounds such mysteries as the legend of Atlantis and the ruins of Stonehenge, and, as logically as we can, sift truth from falsehood and exaggeration.

Early man found himself in a world of unimaginable mysteries: meteors streaking across a star-studded sky, the darkness beyond the campfire's glow, the sound and fury of a volcano's eruption. Our earliest ancestors were probably mysteries to themselves, and totally susceptible to the subjectivity of their world. Fantasies may have been as much a formative influence as toolmaking in the early development of culture. As human beings gathered knowledge and understanding of their surroundings, old mysteries vanished, only to be replaced by others because so much was not understood.

Through the ages—and even today—we ponder over old ruins whose builders left no explanation of themselves and, with little else to go on, imaginations are relentlessly at work. The human mind, inquisitive and curious, is attracted to the mysterious with the same sort of instinct that indicates self-preservation. Perversely, the satisfaction of solving a mystery is only transitory and the mind soon seeks fresh *terra incognita*. In the pursuit of the unknown, the hunger of the mind is so great that it is prone to gullibility, to accept readily the most dramatic explanations of the mysterious, and seemingly to invent mystery if none exists.

A quick glance at our chapter titles will show the kinds of mysteries we selected. There are, of course, many others. Those we included in this book seem to us to be the ones most widely discussed and written about in recent years. We have divided this book into four parts, with a short prologue for each. The purpose of the prologues is to provide continuity, and give the reader a brief idea of what each part deals with. We hope this will be helpful.

We have received much help in preparing this book. We would like to thank the editorial staff of Plenum Press, especially Ellis Rosenberg, Linda Greenspan, and Peter Strupp. Several individuals read parts of the manuscript and offered advice and suggestions. These include John Allen, Tom Burnam, Melvin Hoffman, John Moore, Phil Ostrowski, Lila Kawananakoa, Robert Koll, Bob Funk, George Gill, Nicholas Martin, Edward Brovarski, Ernst Both, and Leonard Smith. We thank also Bob Tilley and Joyce Berry of Oxford University Press.

We are grateful for the technical assistance of Gordon J. Schmahl, Carol McMillan-Feuille, L. W. Griffis, Nancy M. Schaut, Jo Ellen Ruperto, Elmer Berghom, Jr., Kim Fortune, Ann Magida, and Richard Bosher. Special thanks go to Patricia Kay Scott, who assisted throughout the writing of this book. We would also like to acknowledge the valuable assistance of the staff of the Buffalo Museum of Science.

# Contents

INTRODUCTION

# Realms of Reason and Unreason

*Since it is reason which shapes and regulates all other things,
it not ought itself to be left in disorder.*

—EPICTETUS

A Gallup Poll reported by *Newsweek* (June 1978) shows that "Nearly six Americans in ten believe in flying saucers." This translates to about 138 million people. The same survey indicates that 60 million Americans believe in astrology. The elusive "Bigfoot" is thought by more than 30 million people to live on the fringes of civilization. An equal number are certain that the Loch Ness monster exists.

These statistics appear to be borne out by librarians, who are unable to keep up with the enormous public interest in books dealing with these and other allied subjects.

Is there any proof that saucers and monsters exist? That astrology influences our character and personality?

In our own examination of books on these subjects, a major characteristic emerges, and that is that the authors take facts and observations and manipulate them to arrive at apparently new truths. Involved in this process is the use of reasoning and logic, and interpretation of data. There are both correct and incorrect ways to do this. Those who have employed data and reasoning incorrectly to foster offbeat and often unfounded theories have been called *pseudoscientists,* especially by the scientific community. In this chapter we describe scientific thinking and the scientific method, and contrast them with the kind of logic and analysis drawn from what is labeled "pseudoscientific literature." We hope this will serve as background for reading the chapters that follow.

1

## Scientific Thinking

### The Highland Bivouac

The following story is true. It illustrates what science would consider to be evidence leading to the proof of an idea conceived during the gathering of data or information.

Two American archaeologists had been excavating at Tikal, a famous Mayan archaeological site in Guatemala, during the winter of 1959. They decided to take a break from their work and go camping for a few days in British Honduras, which lay to the east. They loaded a Land Rover and set off. They picked the almost uninhabited Mountain Pine Ridge District of central British Honduras to establish their camp and then began leisurely to explore their immediate surroundings.

An open, flat area soon attracted their attention. They sauntered over to it, observing that it was about 100 feet square, and that the grass had been tramped down. Their curiosity aroused, they began poking around and asking each other questions. What was this place? Who had been here before, and when? For what purpose? At first there seemed to be few, if any, clues to help in answering these questions.

Then one man found torn fragments of paper. The scraps contained writing and it was in English, but they could not make out the exact meaning. The mystery deepened. The scientific instincts of the two men held them to the site, patiently probing the matted grass. Clues emerged. Narrow holes in the ground, shrouded in grass, were arranged in a rectangular fashion. Tent stake holes, perhaps? What was puzzling was the presence of shallow dents in the ground, in triangular groupings of three. Inexplicable, it seemed. But then, here were tire tracks, dim but distinct. Not simply single, but also double tracks, suggesting some kind of heavy-duty vehicle. The remains of campfires were discovered, partially concealed in the grass. Finally, maybe luckily, a critical clue was found—a button. It was engraved with what appeared to be a military insignia. With that discovery, the enigma started to dissolve.

The two men reasoned that the square area had been a military bivouac. Because of its size and the areas enclosed by the tent stake holes, the soldiers numbered no more than 20 or 30. They had arrived in heavy, double-wheeled troop transports, set up tents, and staked their rifles tripod fashion in groups of three in front of their tents. This explained the groupings of three dents in the ground, made by the rifle butts. It might suggest also that there were three men to a tent. How long had they stayed? Unknown, but probably more than a day. Possibly several days. Time enough for a button to become lost. Time to eat meals over campfires, and for the comings and goings of booted feet to flatten the long grass. Time enough to write or read letters, tear them up, and scatter the scraps.

Armed with the tentative hypothesis that the square area was a military encampment, the two archaeologists then agreed that a garbage dump should be somewhere in the vicinity. They found it within minutes, just beyond the encampment area. The garbage, consisting mainly of empty cans, indicated that the troops had brought their own food with them, and had not lived off the country to any extent. No remains of game were found, strengthening this assumption.

A few days later the archaeologists verified their theory by consultation with local officials. They were informed that the British government kept small detachments of troops in British Honduras, that they were equipped with heavy vehicles, and that they did conduct maneuvers in the highlands. It was indeed they who had camped for a few days on the spot visited by the archaeologists.

## Scientific Method

In the preceding story, the two archaeologists, perhaps unconsciously, were using the scientific method in their analysis of the military bivouac. They gathered clues without jumping to any final conclusions. When they reached a tentative judgment of what the situation represented, they looked for garbage to test their hypothesis. Later, they sought verification from local authorities. In contrast, a typical pseudoscientific author might have taken the single clue of the military button and constructed an elaborate theory with the button as proof of the presence of a British Army detachment. To the scientific investigator, the button was no more than a single piece of information to be incorporated into an array of data before any conclusion could be reached. Furthermore, the button and the other observations made can only be construed as *evidence* when they are used to prove a point.

Scientific method stands apart from other varieties of thinking because it can be visualized as a group of guideposts and procedures to be used in dealing with problems or mysteries. Generally, scientists abide by these guideposts or rules in their investigations. The step-by-step approach follows this general form:

1. Recognition of a problem or mystery.
2. Formulation of a preliminary hypothesis to explain it.
3. Acquisition of additional facts through observation and experimentation.
4. Formulation of a new hypothesis.
5. Testing of the new hypothesis for its explanatory or predictive value.

Steps 3 and 4 may be repeated many times before we get to step 5, additional facts forcing us to modify our hypothesis, acquiring still more facts, formulating still another hypothesis, and so on. Our "final" hypothesis, in the eyes of the scientist, might still be tentative. Scientists hate to be dogmatic. This is because they use so much inductive reasoning in their work. Inductive reasoning is essentially a matter of the degree of probabilities. For example, if a bee lands on your arm and stings you on three occasions, you might well form the hypothesis that when a bee lands on you, it will sting you. However, there is no guarantee at all that on a fourth occasion the bee will harm you. With repeated stings, your original hypothesis will be strengthened, but only in the degree of probability. Nonetheless, the scientific method offers the best approach to gaining proof of an idea. It is useful in critically examining evidence presented by sensationalist writers. In addition to scientific method, we can use the interesting concept known as Occam's Razor.

## Occam's Razor

Imagine that you are walking along a snowy street. Ahead of you is a man wearing a tall black top hat. To a group of small boys on the opposite side of the street, the hat presents a tempting target. One of the boys throws a well-aimed snowball at the hat. Let us propose two explanations of what happens to the hat. First, at the moment of impact, a band of angels swoops down and invisibly lifts the hat from the man's head. Or second, the snowball itself strikes the hat from the head. Obviously, we chose the latter explanation.

This is a simple illustration of the general scientific "principle of parsimony." Sometimes known as Occam's Razor and named for William of Occam, a 14th-century English philosopher and theologian, it is the principle that the most likely explanation is best. We would add, at least until further evidence can rule out the most likely explanation.

Pseudoscientific authors often violate the principle of Occam's Razor, as we shall point out in the following chapters.

## Fallacies and Their Recognition

**The Importance of Fact.** Facts determine whether statements are true or false. They are supposed to be hard bits of truth—the bricks in a wall of reality. They exist independently, regardless of the opinion of others. The earth is round, and that is a fact. It was still round in pre-Columbian times even though most people thought it was flat. Although facts are regarded as solid, unyielding sorts of things, we all know that facts can be ignored, distorted, embellished, invoked for irrelevant purposes, or otherwise abused so as to impede progress toward truth. If an author who says he or she is clearing up mysteries of the earth abuses facts, then the effort is self-defeating. Abused facts, and the faulty reasoning and conclusions that follow, lead only into what we call the realm of unreason.

In our examination of writings which deal with mysterious events on this planet, we find many authors who take liberties with facts. We drew up a casual list (Table I– 1) showing some of the ways facts and logic can be misused. (We exclude lying and typographical error. We have no reason to believe that any author is a deliberate liar, and of course no books are entirely free from typographical mistakes.) Our list is not exhaustive, nor are the categories mutually exclusive. For example, a false assumption can be seen as factual error if one wishes to do so. Our list is merely a convenient reference for surveying some examples from actual books to illustrate some of these abuses of fact and logic.

**Factual Error.** If a reader spots an error of fact, there is a temptation to be suspicious of everything else the writer says. For example, in his book *Our Haunted Planet,* John A. Keel is talking about Easter Island (p. 57):

> On Easter Sunday, 1722, Dutch admiral Jaakob Rogeveen landed on an island in the Pacific some twenty-two hundred miles from the coast of South America. The first things he saw were hundreds of giant statues squatting near the water line, staring out to sea.

The statues were not "staring out to sea." With few exceptions, they were staring inland. Most of the statues were not in a squatting position, but in a standing position. Trivial? Maybe not. The orientation of the statues no doubt had important implications in the islanders' world view, religion, or other cultural aspects (see Chapter 8). Such errors of fact are a disservice to the reader looking for understanding, and even a disservice to the writer, who wants to be believed.

**Contradiction.** This is closely related to factual error because indeed factual error is involved. A difference is that the reader has an opportunity to see such contradictions and become aware of error even though the reader is

### Table I-1. Some Ways in Which Facts and Logic Can Be Abused

| | |
|---|---|
| Factual error | Speculation |
| Contradiction | Appeal to pity |
| Distortion or exaggeration | Argumentum ad hominem |
| Innuendo | Appeal to authority |
| False assumption | Complex questions |
| Irrelevant data | Hasty generalization |
| Failure to specify | False cause |

not an expert in the subject being discussed. A good example may be found in *No Earthly Explanation* by John Wallace Spencer, who in describing radio-carbon dating (p. 46) states:

> The limit of the radiocarbon clock is no more than forty-thousand years because after that time very little carbon-14 remains in fossils.

So far, so good, until you come to the next page (p. 47) where Spencer notes that ". . . The fossil find was carbon-14 dated as being 135,000 years old." Could this date be a typographical error? We don't know, but at least we know there is an error somewhere.

**Distortion or Exaggeration.** The distortions are most evident on the covers of paperback books. The cover of Elizabeth Nichols's book, *The Devil's Sea,* proclaims:

> Here are terrifying findings about the mysterious, deadly stretch of water where ships, planes, passengers and crews have disappeared without a trace to this day!

There are two distortions here. First, a careful reading of the book shows that a substantial part of it deals with UFOs and other subjects only vaguely related to the Devil's Sea (or, as it is more popularly known, the Bermuda Triangle). Second, at least half of the plane and ship losses did indeed leave "traces" such as wreckage (see Chapter 9).

**Innuendo.** An innuendo implies something that is not necessarily evident from the facts. It is so common an abuse of fact in the sensationalist literature that a single example will suffice. The reader can find many more with only casual research. In Spencer's *Limbo of the Lost,* he tells about the loss of a PB-4YW in 1945, and then observes (p. 19):

> It was as though a small dress rehearsal had been conducted by whatever lurks in the Limbo of the Lost.

This kind of statement would be a tantalizing turn of phrase on a midnight horror show, but not in a book of serious inquiry.

**False Assumption.** Often a writer will make a statement that he regards as true. However, sufficient and accessible evidence exists to cast serious doubt on the validity of the statement. We think of these as false assumptions.

In discussing the Maya civilization, Eric and Craig Umland *(Mystery of the Ancients)* make the false assumption that most scientists consider the Maya to have been a very primitive, unsophisticated people. As they put it (p. 36):

> Can we continue to believe what we are told—that the Maya were a "simple, agricultural people" with no knowledge of technology?

We cannot speak for all scientists, but in cases such as this it is instructive to consult an encyclopedia because it usually offers a view or views that represent a consensus. Here is what the *Britannica* says about the Maya:

The outstanding intellectual achievements of the Maya were in the fields of hieroglyphics, astronomy, and the calendar. . . . In architecture the Maya were far ahead of any other people in the new world. . . . Maya sculpture . . . is now acclaimed as one of the great art styles of the world. . . . Little wood carving has survived, but it is of superb quality. . . . Lapidary work, particularly in jade, is of extremely high quality.

Clearly, the scientific world has great admiration for the Maya and does not consider them as having been a simple, primitive people.

**Irrelevant Data.** Information that has little or no bearing on the question at hand is frequently introduced by popular authors. In her introduction, Adi-Kent Thomas Jeffrey *(The Bermuda Triangle)* assures the reader not to be frightened of the Triangle because the tragedies are, statistically, very small. She then says:

This book is intended to stir up inquiry and answers; not panic and pandemonium.

Her objective, to "stir up inquiry," lies dormant throughout the first 41 pages (or 22%) of the book. These pages are devoted chiefly to the wreck of the *Sea Venture* in 1609 on Bermuda as a result of a storm, the loss of four out of five vessels from a Spanish treasure squadron in 1750, also because of a storm, and other documented history. These first 41 pages, although well written and very entertaining, have little bearing on the stated purpose of the book.

**Failure to Specify.** Here is a quote from *Timeless Earth* by Peter Kolosimo (p. 26):

A human skeleton 17 feet tall has been discovered at Gargayan in the Philippines, and bones of other human creatures over ten feet tall have been found in south-eastern China. . . . In Ceylon explorers have found the remains of creatures who must have been about 13 feet tall, and at Tura in Assam, near the border of East Pakistan (Bangladesh), a human skeleton measuring 11 feet has come to light.

Unfortunately, these reports are not specific. Who discovered these remains? When? Were they deeply buried or near the surface? Are there artifacts associated with them? And most importantly, where are these bones today?

We think these are valid questions. The authors who report such finds do not ask these questions or investigate them, but seem to pass on quickly to other matters. This is strange. To examine, describe, and display such giant human skeletons, if authentic, would be an archaeologist's dream. For a layman, such a discovery would bring fame (and probably fortune). Lack of specification quickly calls into question the validity of the assertions.

**Speculation.** Of all the ways facts can be compromised in sensationalist literature, speculation is most rampant. Actually, a speculative journey can be most fruitful, creative, and rewarding if one undertakes this trip armed with facts. Charles Darwin, for example, arrived at the concept of natural selection after contemplation of (or speculation about) the thousands of observations he had made around the world as geologist–biologist during the three-year voyage of the research ship *Beagle.*

On the other hand, one who sets off upon the sea of speculation armed only with a few scattered facts and a heavy cargo of hearsay and conjecture is like a person cast adrift in a pinnace without a sextant. In many of the writings we have studied, speculation, in order to reach its end point or conclusion, involves some or all of the other abuses of facts we have already considered separately.

To exemplify speculation, let us examine a rather lengthy quote from Eric and Craig Umland, *Mysteries of the Ancients* (p. 103):

> *Doorway to the past.* In the 1950s a story was made public that a map of the far side of the Moon—the side never visible from the Earth—had been discovered carved on a secret, round door in a recently discovered Mayan temple. No further information was ever released on this map; the story has been "forgotten" by scholars and government officials alike.
>
> How could the Maya have been able to make such a map unless they had been on the Moon, or unless they had orbited it in some type of spacecraft?
>
> The Mayan *Book of Gold,* which gives the complete history of the Maya, was hidden when the Conquistadors arrived—hidden in a secret tunnel as yet undiscovered. A temple in Peru bears an ancient inscription that is said to reveal the location of a "tunnel" leading to the Ancient Lost World—a world which is just possibly the Maya's home planet. For, if the Maya had bases on the Moon, they could easily have considered them as their doorways, or tunnels, to the ancient lost world of their original solar system.

The first paragraph is simply a rumor—hearsay. The Umlands do not specify where they obtained this information, nor where this temple is located. Who identified this map? In the 1950s no one had ever seen the far side of the moon, so how could anyone have a basis of comparison for claiming that this is what it was? Note also the innuendo that there may have been a cover-up by scholars and officials. Who and why?

The second paragraph, expressed as a question, shows that the Umlands accept without question the proposition that such a map exists and move forward in their speculative journey to the proposition that the Maya had been to the moon.

In the third paragraph it is stated that the *Book of Gold* is hidden in a tunnel that nobody knows about *(note:* on page 37, the authors say that the book's existence was a legend, and that it was hidden in a temple). Here we have hearsay evidence, acceptance of legend as fact, and contradiction. The authors then shift attention rather suddenly to Peru to mention alleged tunnels leading to a lost world—hearsay—and from this they arrive at the unsubstantiated conclusion that the tunnel leads to the Maya's home planet, assuming they had one. The final assertion, again without evidence, is that the bases on the moon (?) are the tunnels to the Maya's original solar system.

This statement by the Umlands is a fine example of pure speculation. The following abuses were made:

1. Accepting hearsay as evidence
2. Failure to specify
3. Contradiction
4. Acceptance of legend as fact
5. Unsubstantiated conclusion
6. Innuendo
7. Irrelevant conclusion

There is not one substantiated fact offered in the entire passage. If one allowed oneself to be amused, it would be at the realization that there are no errors of fact in the entire passage because the entire passage contains no facts.

**Appeal to Pity.** This kind of approach usually depicts the author as pitting himself against the "establishment" of orthodox science. From the Umland's introduction, *Mysteries of the Ancients* (p. 9):

> . . . we have been forced to attack some of the "sacred cows" of modern archaeologists, anthropologists, and historians. We realize that we shall be criticized for doing so. In the search for truth, however, sacred cows must often be sacrificed.

The authors will be criticized by scientists for searching for the truth.

**Argumentum ad Hominem.** This type of argument is an attack on an individual rather than what he says, and often takes the form of ridicule and sarcasm. In his book *The Gold of the Gods,* von Däniken raises the question (pp. 10 and 12):

> Is man really prepared to admit that the history of his origin was entirely different from the one which is instilled into him in the form of a pious fairy story? Are prehistorians really seeking the unvarnished truth without prejudice and partiality?

These questions are an attack on the honesty of individuals rather than on what they say and document about man's origins. Von Däniken continues this theme a bit later (p. 34):

> I should like to know what tricks scholars will use to displace this fabulous metal treasure of inestimable archaeological and historical value. . . .

Here is another attack on the honesty of scholars, generalized (apparently) to include all of them, and predicting that the scholars will be dishonest in the future.

**Appeal to Authority.** Certain arguments can be buttressed by calling upon an authority or expert. Many writers do this—all well and good. However, one must be careful to note if the expert's testimony *coincides* with his field of expertise.

In an interview of infantry Captain John Alexander by author Brad Steiger *(Mysteries of Time and Space),* Steiger questions the captain about an underwater archaeological (?) site near Bimini which some regard as a part of the lost continent of Atlantis. In this interview, the captain is the "expert." His credentials are given as follows (pp. 55ff):

1. Alexander has a military background.
2. He is Silva Mind Control representative for Hawaii.
3. He has been around the world a couple of times.
4. He has studied Buddhism.
5. He is an experienced diver.
6. He is an underwater demolition expert.

This is a fine set of credentials if one wishes to interview the captain about infantry tactics or underwater demolition. On those topics we are reasonably certain his answers would be authoritative. The captain was not being questioned about these matters, but rather about his opinion of the presumed archaeological site he had observed during one or more dives. He was questioned (in part) on the following points:

1. Was what he saw at Bimini a small section of what was once a much larger city?
2. What was the composition of the stones?
3. Are the structures off Bimini the remnants of a precataclysmic civilization? (The captain said yes.)
4. Did he find any artifacts in addition to the structures?
5. How advanced were they? Our equals? (The captain was affirmative.)

The captain's responses, some at great length, show that he is neither an archaeologist nor a geologist, and his testimony is unfortunately useless on these points.

**Complex Questions.** Many books ask numerous complex questions. Indeed there would not be much left of some books if the questions were removed. These questions are posed in such a way as to presume that previous questions have been asked and answered satisfactorily. Richard E. Mooney *(Colony: Earth)* is a source of many examples. Here is one (p. 17):

> if man spent thirty to thirty-five thousand years as a wild hunting nomad, why did he suddenly become civilized?

This two-part question ("if" and "why") is posed as though the answer to each part, given as propositions, is "yes." Questions such as this serve no purpose other than to confuse the reader. On the same page, Mooney also asks, "How and why was the Great Pyramid constructed?" This is a more meaningful question, for which archaeology has some answers and seeks others. However, when such a question, without being answered, is imbedded in the context of irrelevant questions, it takes on the "taint" or innuendo of the irrelevant question.

**Hasty Generalization.** We all sometimes make generalizations, especially in everyday conversations. However, they should not appear blatantly in a book that presumably has been examined by several knowledgeable reviewers. John A. Keel, for instance, in discussing the ideas of Velikovsky, notes (*Our Haunted Planet*, p. 79) that these ideas were not well received by the scientific community. He then says of scientists:

> They resented the fact that a psychiatrist dared to speculate on astronomy and archaeology. He was an intruder. Above all they resented the fact that his book was very well written (most scientists are miserable writers).

Author Keel generalized the nature of the resentment of the scientific community as being due to (1) the fact that Velikovsky is a psychiatrist, and (2) the belief that Velikovsky's book was "very well written." We do not know how many scientists even *knew* that Velikovsky is a psychiatrist. Does author Keel? Whether the book was very well written is a question people can judge for themselves. A well-written book may be full of mistakes. Many scientists think Velikovsky is a fine writer, but do not admire his selective pruning of those facts that militate against his preconceived ideas (see Chapter 13). Keel's final generalization that "most scientists are miserable writers" has no foundation.

**False Cause.** The confusion of cause and effect is a commonplace with which we are all familiar. You have a headache and take an aspirin. The headache goes away. Was the aspirin the cause? Not necessarily. The headache might have gone away anyway. It is difficult to determine without much experimenting. It is a bit easier to pinpoint false causes in the great mysteries literature. Here is an example from *The Mystery of Atlantis* by Charles Berlitz; speaking of the breaking of the transatlantic cable in 1898 about 500 miles north of the Azores (Atlantis "territory"), Berlitz says:

> While the cable was being searched for, the sea floor in this area was found to be composed of rough peaks, pinnacles and deep valleys more reminiscent of land than of sea bottom.

Berlitz seems to be saying that the varied submarine topography found is more likely to have evolved as land above sea level and is now submerged. In response, we would point out that, especially since World War II, sea-bottom topography has been mapped extensively. Contrary to the old-fashioned visualization of the ocean floor as a featureless plain, the sea bottom around the world displays varied topography due to the operation of such well-recognized activities as erosive turbidity currents, submarine volcanism, and faulting. The features Berlitz describes are quite distinct from landforms, which are shaped by somewhat different processes, especially stream erosion and chemical weathering.

Another false cause appears in the same passage of Berlitz's book (p. 89) when he states:

> Grappling irons brought up rock specimens from a depth of 1700 fathoms which proved, upon examination, to be tachylyte—a vitreous basaltic lava which cools *above water* under atmospheric pressure [italics in the original].

To the logical layman, this would seem convincing proof that sea bottom 1700 fathoms (10,200 feet) deep had once been above water, a persuasive "plus" for Atlantis. However, it is another false cause. The *cause* that determines the evolvement of the rock type in question is the *rate of cooling,* not whether it was cooled above water. These rocks can form above or below water and under pressures other than atmospheric pressure (see Chapter 10).

Finally, we come to the question implicit in this chapter. Is pseudoscience harmful? There are two opposing views of the long-term effects of pseudoscience on society. On the one hand, some writers suggest that pseudoscience is a form of irrationality that may be dangerous and costly to society. This is the position taken by the American Humanist Association and its daughter organization, the Committee for the Scientific Investigation of Claims of the Paranormal. Both have taken a strong public stand against astrology and publish a journal to promulgate their crusade against pseudoscience.

Physicist James Trefil considers it an overreaction. He writes (*"A Consumer's Guide to Pseudoscience"*):

> Pseudoscience has been around at least as long as (and perhaps even longer than) conventional science. Perhaps it serves some deep need of human beings to believe that there is still some mystery—something unknown—left in life. Maybe the unknown thrives because people like to see the pompous scientific establishment discomfited ("Okay, Mr. Know-it-all, explain *this* one."). Or maybe it's just that P. T. Barnum was right about a sucker being born every minute. None of these interpretations constitutes a threat to conventional science. . . . At its worst, pseudoscience is a minor inconvenience of the cocktail party variety; and at its best, it is good entertainment.

Our own appraisal of sensationalistic pseudoscience covers a broad panorama. And we firmly believe that pseudoscience truly is an old stage with new actors appearing down through the years. But although it may be a natural condition of life, we do not feel that the collective world of pseudoscience can be dismissed as no more than a minor inconvenience. Take the individual case of archaeology. It is first of all a science that tells us about ourselves, from all times and places. It enriches our lives. But more than this, it is a major source of information on the background of our history, our diseases, past climates, and natural disasters.

To accomplish these things, archaeology depends almost entirely on public support. Yet when the adventures of Erich von Däniken are identified as archaeology, the public is misinformed. When films and magazine articles make selective use of evidence to argue persuasively that archaeologists have somehow wrongly reconstructed our past, the public is again misinformed. The vestiges of human prehistory are rapidly disappearing under the impact of 20th-century technology. Pseudoscience may be more than a minor inconvenience when fantasy competes with scientific empirical research.

This completes our introductory survey. It has not been our purpose to be supercritical of any of the authors we have cited; rather, our purpose has been to point out to the reader examples of the mistakes we see, and make the reader aware of them. In most of the following chapters in this book, we will be looking closely at specific subjects to see how well they stand up to the principles and concepts of reason and logic.

# PART I
# DIFFUSION AND SUPERDIFFUSION

# Prologue

When we take a broad look at much of what has been written about man and his history, we find an underlying theme always emerging: the complex matter of human migration.

Science has pursued the fossil evidence of human origins, particularly in East Africa where evidence seems to indicate that man may have originated more than 2 million years ago. From the beginning, migratory man moved from one place to another and eventually from continent to continent. He thus began making a cultural imprint on the world. Striking parallels, especially between continents, may be the result of cultural diffusion involving migration (contact and imitation) or the result of independent invention. Scientists, probing pre-Columbian levels of a Pennsylvania rockshelter or the submerged wreckage of an ancient galley in the Mediterranean, continue to ask the five Ws of the newspaper reporter: the who, what, when, where, and why of man's movements and the spread of his culture.

It is no easy task to retrace all the countless and circuitous bypaths of worldwide diffusion through the many millennia of man's history. Clues are meager and interpretation must be guarded. Although written records have existed for some 5000 years, critical problems remain.

Our main purpose in Part I is to examine the subjects of migration and diffusion because these ideas constitute a major theme in what we consider to be sensationalist literature. At first glance, there might seem to be little connection between the space "chariots" of Erich von Däniken and arguments over who built certain stone structures at Mystery Hill, New Hampshire; however, both are diffusionist questions.

Chapter 1 is introductory, giving the reader an overview of the history of ideas on diffusion, particularly as it relates to the discovery of the New World. We see these ideas as recurring in popularity cycles. The nature of the archaeologist's work, and his understanding of diffusion and migration, are discussed, along with pertinent examples.

In Chapter 2, we contrast some of the methods, data, and attitudes of the scientists with those of popular authors who seek easy, pat answers to very complex questions. Mystery Hill in New Hampshire serves as a detailed case history to illustrate this contrast.

Chapters 3 and 4 deal with the "ancient astronaut" hypothesis, which, as pointed out in Chapter 1, has a historical continuity with other diffusionist ideas, but is couched in terms familiar to contemporary society. The basic concepts of von Däniken and his fellow ancient astronaut authors are summarized in Chapter 3. Chapter 4 is devoted to a critical response to the methods, data, logic, and reasoning of these authors.

In our space age world, the UFO or unidentified flying object often is conceived by many people to be a spaceship from another civilized planet. The connection with the ancient astronaut hypothesis is obvious, and is looked upon by us as contributory to a superdiffusionist theme. Chapter 5 examines the subject of UFOs, with a critical look at the views of scientists as well as those of writers favoring an extraterrestrial hypothesis. Diffusion, or the spread of culture in the course of human history on earth, becomes exaggerated to super proportions by the advent of modern, especially space, technology.

# CHAPTER 1

# From Continent to Continent
## Making Ends Meet

*Discussions of diffusion are apt to degenerate into combats wherein only dust is diffused or else to ascend into an ether so diffuse that the disputants are left balancing probabilities inherently unsusceptible of statistical treatment.*

— GORDON CHILDE

## Journalist's Rationale

"Celtic Marks Found in New England," says the newspaper headline. The article continues by stating that Celtic peoples of western Europe were living in the region of New Hampshire and Vermont more than 2000 years ago. Proof comes from rock inscriptions in an ancient language, Ogam, deciphered by members of the Epigraphic Society.

Could this really be true? Will the early history of North America have to be rewritten, this time with Europeans beating Columbus to America by 1500 years?

One of the realities of successful journalism is that items of marginal news significance need not adhere to the traditional rule of objectivity. The fact that many archaeologists, historians, and epigraphers might not agree with the findings of alleged Celtic remains will not change the news item. In the journalist's idiom, the romanticism of lost continents, UFOs, ancient astronauts, and epic pre-Columbian (pre-Christopher Columbus) voyages between continents has the impact of novelty and the news value to help sales. A carelessness in reporting that misleads the public cannot be tolerated if it concerns health, taxes, or some other potentially harmful area. Unfortunately, the entertainment value of many genuine mysteries often leads journalists and freelance writers to say, in effect, "So what if the public believes?" The mass circulation of speculation, disguised as evidence, is a kind of brash illogic that scientists detest and that the public seems to demand.

Rather than attack press irresponsibility, we might recognize that such sensationalist journalism will automatically arouse suspicions among professional scientists. It does so for three major reasons:

1. The subject matter tends to follow cycles of popularity and to be strongly influenced by contemporary beliefs.

2. Fanciful reconstructions presented by mass-market journalism become fixed in the public mind, often in the face of contrary evidence—evidence that is compelling to most of the scientific community but not to the public mind.

3. Authors whose universe is the "mysterious" use archaeological data and yet typically they do so without basic knowledge of the methods of that science.

## Popularity Cycles

Most of us are aware of ebb and flow in our society—some things are "in," other things, "out." Examples are yoga, men's ponytail hairdos, mink coats, and bathroom telephones. Depending on time and place, these and other symbols are, or are not, in vogue. We see this same pattern within the realm of man's inquiry into his own past.

The current belief that our planet was visited by ancient astronauts, for example, would seem to suggest that we have a modern movement, not to be thought of as fantasy but rather as important "new" evidence. Do we not have here simply the latest popular idea to explain diffusion, the spread of culture? Underlying the ancient astronaut concept, and essential to it, is diffusion. A convenient mechanism, spacemen, is used to explain many ostensibly similar cultural manifestations to be found from continent to continent. Perhaps the ancient astronaut hypothesis is merely the most recent and popular "in" way to explain cultural diffusion. It is even considered by some that motion pictures exploiting outer space themes are another form of diffusionist literature. A well-known example was *War of the Worlds,* derived from H. G. Wells's book of the same name. *2001: A Space Odyssey* and *Close Encounters of the Third Kind* came on the scene with fresh impact during the 1970s. Perhaps it is too early to know if outer space themes will have cyclical popularity in the same way as American Indian origin theories.

On the subject of American Indian origins, Robert Wauchope, noting the popularity cycle of various theories, points out in *Lost Tribes and Sunken Continents* that the Ten Lost Tribes of Israel were once widely believed to be the remote ancestors of the American Indians (p. 3):

> . . . for in those days ancient Hebrew ethnology as described in the Old Testament was about the only well-documented "primitive" way of life known and therefore the first to occur to a seeker of Indian relationships.

Wauchope is speaking here of the 16th century. Columbus had recently discovered not only America but American Indians, whose presence there could not be immediately explained. Who were these natives? How did they happen to arrive in America? From what known people had they descended? At that time in Europe it was almost universally believed that all men were descended primarily from Adam and Eve and from Noah and the survivors of the flood. The theological nature of these explanations was a product of the times when origin theories had to be made consistent with the Bible.

Later, in the 18th and early 19th centuries but prior to the development of prehistoric archaeology, England and the continent developed a taste and appreciation for the achievements of ancient Mediterranean civilizations. The maritime empire of Phoenicia was especially admired. Living in what is now Lebanon with the chief colony at Carthage, the Phoenicians were famous for their seafaring skills and commercial enterprise, their wealth augmented by the founding of trading posts throughout the Mediterranean and down the west coast of Africa. No wonder scholars favored the idea of Phoenician expeditions to account for American cultures. Thus the 18th and early 19th centuries had introduced new interpretations.

Then a new wave of popularity appeared: Egyptian diffusion. About the middle 1800s, prehistoric archaeology itself began to show signs of birth as a recognized branch of learning. Then, as now, Egyptian civilization and especially the pyramids were great objects of wonder. Much earlier the Greeks had been fascinated by Egyptian ruins, but Greek knowledge of the great civilization of the Nile was very poor. Long after the Greeks, but prior to the development of archaeology, Napoleon made Europe aware of the importance of Egyptian antiquities following his arrival in Egypt in 1789. Napoleon and his French scholars acquired important archaeological materials, including the famous Rosetta Stone which was deciphered and found to be the key to Egyptian written records. But by the 19th century the ancient Egyptians were becoming the object of systematic archaeological investigations to bring to light the lives of these fascinating people. While "dirt archaeologists" were starting to reveal the long dynastic history of Egypt, other explorer–writers were finding, in the lowland forest of Mexico and Guatemala, the extraordinary evidence in sculptural art and architecture of ancient New World civilizations.

We are not surprised then to find that popular interpretations shifted from biblical and classical to Egyptian solutions as the most likely explanation for American Indian cultures. The power and organization of the Egyptian state even led some writers to conclude that practically all cultures were recipients of a complex of traits originated in Egypt and carried throughout the world by sun-worshiping peoples of that country. We will refer to that theory again later in the chapter.

Next in order came the theory that natives of the South Pacific islands were in reality American Indians, maritime arrivals from both Peru and the northwestern coast of North America. This theory, inspired by Thor Heyerdahl and all but sanctified by the daring and dangerous raft voyage of the *Kon-Tiki,* did much to stimulate Pacific island archaeology, both for scholars concerned with Oceanic cultural history and for those who eagerly accepted the notion of definite relationships between Polynesian and American cultures.

Thus, we come full circle to the doctrine of visitations from outer space, perhaps the most widely disseminated and controversial hypothesis of human history. Using various objects and drawings from the prehistoric past, ancient astronaut writers point to outer space contact with American Indians. In Chapters 3 and 4 we will examine the theories and evidence of the ancient astronaut concept in more detail. Here we simply say that outer space theorists take their place on a long-lived list of speculative histories of man. Much depends on the age and its circumstances. The spectrum of explanations, from Adam and Eve to ancient astronauts, spans a long period in the growth of human knowledge. Each theory owed its popularity to the fact that it could be easily understood by a society viewing the evidence in terms of its own scientific knowledge.

The paradox in this is that, imperfect as it is, our 20th-century knowledge of the world and our knowledge of human history and the gradual emergence of civilization are, in a figurative sense, light years ahead of that of earlier centuries. Yet regarding American Indian origins, very few of the old ideas become outworn, discarded notions. Writers can now embrace all the older ideas plus the extraterrestrial hypothesis, which we argue is a remnant of the past expressed in the terms of today's consciousness of space science. Writing in the early 1960s, almost anticipating some of the ideas of von Däniken and his followers, Wauchope had this to say in *Lost Tribes and Sunken Continents* (p. 5):

> An ever increasing number of people in this country and abroad are intrigued, in some cases compulsively obsessed, with all forms of symbolism, especially if it is somehow esoteric, and best of all if it can be related to the culture and origins of the American Indian, particularly the Maya, Aztec, and Inca, to whose colorful prehistoric civilizations they are drawn like flies to honey. Here they can delve spellbound in bizarre symbols, the most dramatic and sacred rituals, mysterious hieroglyphs, jungle-shrouded

ancient temples, non-Indo-European languages whose strange syllables lend themselves to endless games of linguistic blind-man's bluff, and polytheistic religions with strong mathematical, astrological, sadistic, and phallic overtones. Their fascination with these becomes an addiction, in some cases literally a religion, and in holding fast to their beliefs about aboriginal America, in the face of often brusque rebuffs from professional anthropologists, they feel persecuted, martyred to a sort of semi-scientific, semi-religious destiny that must not be denied. Their writings were frequently larded with references to the Almighty, theology, and ethics in passages where the uninitiated reader fails to see the relevance. Dealers in used books are well aware of them, for they haunt the sections in bookshops marked "Esoteric, Occult, and Curiosa."

## The Ancient Mariners

Let us take a look at some other literature on early transoceanic crossings. The web of arrows shown in Figure 1-1 indicates only a partial list of proposed human links from Asia and Europe to America in prehistoric times. The range

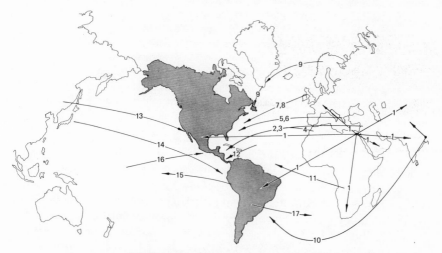

AMERICAN INDIAN ORIGIN THEORIES

| | | |
|---|---|---|
| 1 Egypt–Heliocentric | 7 Irish | 13 Fu Sang |
| 2 Lost Tribes of Israel | 8 Welsh | 14 Jomon-Valdivia |
| 3 Phoenicians | 9 Vikings | 15 Kon-Tiki |
| 4 Ra | 10 Hindus | 16 Mu |
| 5 Greeks | 11 Africans | 17 Origin–America |
| 6 Romans | 12 Atlantis | |

Fig. 1-1. A network of alleged interhemispheric voyages in prehistoric times.

of choices reads like a Mardi Gras extravaganza: Hindus, Chinese, Dutch, Poles, Basques, Egyptians, Romans, Koreans, Hebrews, Phoenicians, Turks, Japanese, Assyrians, and others. One author even proposes that they all discovered America. Revival of these theories from time to time is likely to continue. A good story is good not just once, but many times over.

Take the example of the Fu Sang hypothesis. The story was first told in early Chinese literature of a seafaring expedition in A.D. 458. The voyage was made by five Buddhist monks from China to a faraway land known as Fu Sang, a place that, in the minds of some, can be identified as pre-Columbian Mexico. Fu Sang became part of the firmly entrenched lore after 1761 when a Frenchman named de Guines published his opinion in support of the theory. As Professor Wauchope and others have observed, Fu Sang and similar theories were seriously and emotionally debated through the 19th century. The following quotation from Frederic de Hellwald, speaking at the International Congress of Americanists at Nancy, France, in 1875, not only condemns the lack of evidence but also reflects an impatience with the continuing reappearance of the Fu Sang beliefs (Wauchope, p. 102):

> This Fu-Sang legend keeps turning up periodically as obstinately and as regularly as the Sea Serpent apparition is reported in our journals. Just as no one has ever admitted personally to having studied that animal zoologically, so has no one ever scientifically proved the discovery of America by the Chinese. Dr. Bretschneider several years ago amply refuted this fable, which didn't prevent an English book on it from appearing recently. It is to be feared that the refutation of M. de Rosny and Lucian Adam will not put a stop to the reappearance of the monster. The Congress of Nancy would render a true service to science in declaring that it holds the Fu-Sang theory to be a scientific sea serpent and in forbidding it to infest henceforth the latitudes of Americanism.

As predicted, the speaker's prophetic words were answered when, almost a century later, in 1973, a research proposal was submitted to the Office of the Vice-Chancellor of Academic Programs of the State University of New York in Albany. Funds were requested for a projected expedition intended to duplicate the Fu Sang voyage—not just to test the feasibility of such a voyage but to prove the theory. Fu Sang was the subject of another book, *Pale Ink,* by Henriette Mertz, published first in 1953 and reissued in 1972 under the title *Gods from the Far East: How the Chinese Discovered America.*

## What Do Archaeologists Think?

Up to this point we have attempted merely to show a historical continuity of popular ideas and judgments about man's origins and movements in the world. Let us turn once more to the difference between professional scientists and popular writers. The contrast shows up in the tendency of some popular writers to conceive easy answers to very complex archaeological problems, ignoring real investigative methods. Let us look a little more closely at archaeological science.

Archaeology is a branch of anthropology. Anthropology, the "science of man," is concerned with culture and its enormous range of phenomena: art, religion, myth, political organization, agriculture, dance, warfare, economics, personality, kinship, technology, and so on. Moreover, these and many other subjects are studied by the anthropologist in the contexts of simpler, nonliterate societies, as well as in more complex, urban settings, both in the past and the present. Anthropology is a broad and complex study. Archaeology is a special investigation of the prehistoric span of man's history. It depends on the often fragmentary remains of human activity in the past. Since archaeology is an area of specialization within anthropology, the archaeologist is also an anthropologist. This means that his education includes not only training in the application of science to excavation and general field methods, but also a solid grounding in theoretical concepts essential to the understanding of human history. What are the theoretical concepts? In the course of sorting through 400 years of different ideas concerning prehistoric contact between peoples, concepts such as migration, convergent evolution, psychic unity, diffusion, and invention were important elements in debate, even though they were not always identified as important issues. Most characteristic perhaps are the opposing arguments of diffusion and independent invention.

## Diffusion and Independent Invention

Diffusion is one way in which cultures have changed or evolved. In the simplest definition, diffusion refers to a spreading out or distribution by contact. One way of grasping the idea of diffusion is to compare it to the origin in Asia or Africa of an influenza virus which then moves or diffuses across land masses and oceans alike. However, that is not a scientifically useful comparison. To an anthropologist the process of diffusion involves aspects of human culture: the spread of artifacts, or of religious and social traits. For example, Christianity, following its birth in the Near East almost 2000 years ago, is now found among such widely diverse groups as Eskimos, Australian aborigines, and tribes of the South American rain forest. But neither is Christianity a good example, because the attempt to spread its message throughout the world was planned, organized, deliberate, and militant. One might call it forced diffusion.

Let us take still another example, the couvade—a classic in anthropological literature. Couvade is the curious practice at the time of childbirth when the father retires to his bed, receives visitors, observes certain taboos, and, in general, symbolically imitates the role of the mother. The custom has been described for selected times and places around the world. Couvade was found to be practiced on the Mediterranean island of Corsica in the first century A.D. Marco Polo reported it in China. Elsewhere it was found among the Basques of Europe, some South American Indian tribes, and in North America among California Indian tribes. Did this cultural trait originate in one place and later disperse from tribe to tribe by diffusion? Such was the argument of the diffusionists. Or could couvade have been invented more than once?

Independent invention, or the theory of parallel development, is precisely the opposite view to diffusion. When similar or identical traits, or even a group of traits, are found in two parts of the world with no known history of contact between them, the traits can be explained as having been invented in various centers, independently. On the subject of couvade the diffusionists were opposed by the inventionists, who said that instead of having a single origin, couvade had been invented independently in several major areas of the world and from these it diffused locally.

Perhaps the best-known case study of the extreme diffusionist position from the past was the Pan-Egyptian or Heliolithic Culture Theory (alluded to earlier in this chapter), which is to be found in the writings of G. Elliot Smith, *The Migrations of Early Culture* (1923), and W. J. Perry, *The Children of the Sun* (1923). The hypothesis stated that Egyptian culture, identified by associated traits of mummification, the carving of stone statues, sun worship, megalithic building construction, and other activities, spread from Egypt throughout the world. The theory was effectively criticized in an early study by Roland Dixon, professor of anthropology at Harvard University. In *The Building of Cultures* (1928), Dixon illustrated the dangers of the casual use of the idea of diffusion by showing that such traits as the worship of the sun and the carving of stone images were only superficially similar and that the indiscriminate matching of such traits between continents is therefore misleading. Dixon summed it up with these words (p. 263):

> Throughout, one finds statements for which there is no foundation, coupled with the neglect of pertinent and easily ascertainable facts, and a reasoning which shows a fantasy of imagination ill suited to the elucidation of the very intricate and puzzling problem of culture origins and growth.

This statement might well apply equally to much of the bookstand literature we see today, although Dixon wrote the words more than 50 years ago.


## Migration

Migration is a word that is misused freely by writers who carelessly offer it as an explanation for similar cultural phenomena perceived around the world. Such writers usually make no distinction between migration and diffusion.

Diffusion implies the taking over of traits by imitation. These traits may diffuse from group to group, in the form of ideas or objects. The bow and arrow may have made their way around the world largely by this means.

Migration, on the other hand, refers to the wholesale movement of people breaking away from their original settlement, carrying their culture and readapting it to the conditions of their new place of residence. An interesting example is the Polynesian migration to New Zealand at a time when geography itself was a great mystery. How and why did a human migration take place on a seaway of more than 1000 miles along which land could seldom be seen? When did settlers arrive and from where? Archaeological investigations point to the Society and Marquesas Islands of tropical eastern Polynesia as the most probable source of the immigrants (Fig. 1-2) in that artifacts of the first settlers are closely related to those found in the home islands of eastern Polynesia. New Zealand lies far to the south of other Pacific islands and has colder weather. Meticulous archaeological studies in New Zealand have shown the success of those primary settlers in adapting from tropical to temperate climate, with new plants to be cultivated, new tools and weapons to be made from new materials, and so forth. Radiocarbon dates from both ends of the migration trail suggest that the first Polynesian population move took place in the eighth or ninth century A.D. The question of why such a migration took place is a part of the

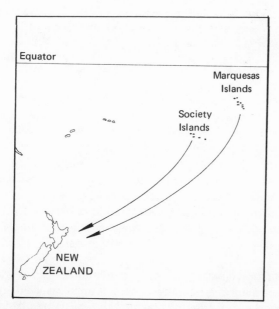

Fig. 1-2. Map of south Pacific Ocean showing probable route of maritime migration to New Zealand from tropical Polynesia.

larger issue of general Polynesian origins and dispersals, and for the moment it remains unanswered.

The New Zealand example illustrates that archaeologists, adhering to scientific method, are particularly critical of their own evidence for migration. They make close and reliable estimates of age. They test buried prehistoric horizons for possible local sources of alleged imports. Culture traits brought by migrating people must be shown to be new, in other words, without local prototypes. Furthermore, not only must the migrational homeland be identified, but questions such as why and by what routes the migrants traveled should be asked and if possible answered. Archaeologists who take a cautionary and objective view of migration theory are appalled by the ready and facile use of migration as a solution for the unexplained. For instance, a proposed global migration (or migrations) is described in the concluding sentences of *Megalithic Software* by L. B. Borst and B. M. Borst (1975). Speaking of the spread of mathematical knowledge, they state (p. 184):

> We have no way of knowing how many sites we have missed. We therefore reject bodies of data which we cannot understand or which would seem to us unprofitable, and retain only those data which are potentially understandable with available time and knowledge. Among this residue we find a story, unexpected and provocative, of prehistoric mathematics and of a cultural migration, perhaps representing a religion, extending far beyond the bounds of our imagination; from Ireland to Japan; from the arctic circle at least to Egypt. This culture precedes historical records even in Babylon and Egypt. The origin of the culture cannot yet be identified. The time of the migration to Japan is unknown.

Even without knowledge of the where, when, how, or why of the alleged culture, the authors perceive a far-flung migration. There is no marshaling of evidence of migration; no mention of intermediary, third-party cultures passing along ideas through contact; no thought of possible multiple origins. Instead, the word "migration" is used as a convenient and easy answer to an enormously complex problem.

Of course we cannot deny that during man's existence both diffusion and migration have taken place. Throughout history human beings have migrated. They left their ancestral homes, either voluntarily or involuntarily, but with the end result of occupying most of the livable parts of the world. Cultures grew. Civilizations emerged with central governments. Whether simple or complex, societies expanded by borrowing or perhaps by invasion, or by other diffusionary means.

In this chapter we have seen that the question of Old World sources of native American cultures is not a new one. It is an unresolved mystery that is constantly regenerated by the remarkable likenesses found between the Old and New Worlds. Where does anthropology stand on the subject of these correspondences? Most anthropologists believe that pre-Columbian New World cultures are American, New World developments. A point to be remembered is that solid evidence accruing from radiocarbon dates and artifact material collected during hundreds of field studies shows that man arrived from Asia, via the Aleutian–Bering Strait route, perhaps as many as 30,000 years ago, when the worldwide sea level was low and a land bridge existed. The earliest arrivals brought their culture in the form of simple tools and weapons. Building on that meager cultural heritage, prehistoric American Indians progressed, some more than others, toward high levels of cultural achievement. Many years of careful scholarship have convinced us that the main course of American cultural development is home grown. Furthermore, cultural progression in prehistoric America roughly paralleled that of Europe. We are by no means alone in making this observation. This is the story as archaeologist A. V. Kidder has put it in *Expedition* (pp. 22–23):

> . . . the analogies between the history of pre-Columbian America and that of the Old World must again be stressed. In both hemispheres man started from cultural scratch, as a nomadic hunter, a user of stone tools, a paleolithic savage. In both he spread over great continents and shaped his life to cope with every sort of environment. Then, in both hemispheres, wild plants were brought under cultivation; population increased; concentrations of people brought elaboration of social groupings and rapid progress in the arts. Pottery came into use, fibers and wool were woven into cloth, animals were domesticated, metal working began—first in gold and copper, then in the harder alloy bronze. Systems of writing were evolved.
>
> Not only in material things do the parallels hold. In the New World as well as the Old, priesthoods grew and, allying themselves with temporal powers, or becoming rulers in their own right, reared to their gods vast temples adorned with painting and sculpture. The priests and chiefs provided for themselves elaborate tombs richly stocked for the future life. In political history it is the same. In both hemispheres group joined with group to form tribes; coalitions and conquests brought preeminence; empires grew and assumed the paraphernalia of glory.

Professor Kidder's reconstruction does not mean that other visitors could not have crossed the oceans to reach America's shores during our prehistoric past. Older, isolationist views may have to be modified. For example, after many years of nonacceptance by scientists of claims for Vikings in the New World, the first proven remains of a Norse settlement in the Americas were discovered on the northernmost tip of Newfoundland. What had been speculation became fact, namely, that Norsemen sailed to North America about A.D. 1000.

And there are other interesting possibilities of prehistoric intercontinental contact. Our common sense tells us that the Asian seafarers who settled the far distant islands of Polynesia had the sailing skill to continue on to America.

Perhaps they did. It is not a simple matter to put such a proposition to the test. Scientists are creatures of fact, and archaeology is a science dependent on cultural bits and pieces, the significance of which may come much later. Yet America is still a great field for new discoveries.

Our subject, diffusion and superdiffusion, is, at the same time, an old and a new problem. In our next chapter, we will examine some neodiffusionist views and the kind of data that are frequently offered as evidence. The reader should remember that we are speaking from the point of view of science, governed by the procedures and methods discussed in the Introduction.

**CHAPTER 2**

# Stones, Suppositions, and Science

*Superstitions, like bats, fly best in twilight.*

—BACON

## The Asian Invasion

Man is perhaps the most widely distributed species on earth. We know he made his way to the fertile American continent only after his earlier history was laid down in Africa, Europe, and Asia.

As soon as it was known, about 1750, that North America and Asia are nearly joined at the Bering Strait, Europeans began to theorize that this was the pathway leading to the first colonization of North America. A mere 60 miles of cold Bering Sea water separate the shores of Siberia and Alaska.

During the last and most widespread advance of continental ice sheets across the northern hemisphere, the worldwide sea level fell at least 300 feet. Beginning about 10,000 years ago, the ice sheets retreated and the meltwaters caused sea level to rise again. Before this event, the Bering land bridge had existed for several millennia, providing access for nomadic migrants to the North American continent.

The first people were hunters and gatherers, and they probably entered America while following the herds of large Pleistocene game animals such as the giant bison, mammoth, and other mammalian species that were part of their food supply. By 12,000 B.C. they were well adapted to hunting the herds. Archaeologists call these people Paleo-Indians.

It is a long reach back for us to know much about these people, but fortunately they left behind widely scattered evidence of their hunting activity in

29

Fig. 2-1. A fluted projectile point of the type used by
early North American hunters.

the form of stone tools. Their most characteristic weapon was the fluted
projectile point. Figure 2-1 shows an example of these flaked spear points with
a long shallow groove, or "flute," removed from each side. Archaeologists
recognize a Paleo-Indian site wherever these distinctive spearheads are found,
from subarctic Canada to South America.

The nomadic Paleo-Indian hunters probably gathered wild plant foods in
addition to hunting, and, in any case, they practiced their hunting-gathering
livelihood in an environment that was gradually warming as Pleistocene ice
diminished. Ten thousand years ago, North America was physiographically not
too different from today. The same mountains bordered the same plains; there
were spring-fed streams, deserts, and valleys.

It was *Homo sapiens*—fully developed modern man—who moved south
from Alaska. At first they made simple camps along their way toward populat-
ing the Americas. The first arrivals were probably followed by later waves of
Asian newcomers. After all, the New World had plenty of room for immigrants.
For thousands of years then, the native American Indians were busy adapting
their cultures and societies to the rich and varied environment they found in
America. Thorough and painstaking studies in archaeology have shown us that
certain major steps were taken during that long period of prehistory:

- Campsites of the early hunting and gathering tradition changed to
  settled village life.
- Population increased.
- Indian corn (maize) and other agricultural crops, such as squash, beans,
  melons, tomatoes, were cultivated.
- Pottery was invented.

In countless other ways, Indian cultures, customs, and languages began to take on distinctly American features. Then certain inheritors of the early American Indian tradition went even further. The Aztec, Maya, and Inca carried culture to a higher level of civilization (Fig. 2-2). They excelled in the minor arts of pottery, textiles, jade, and goldwork, and they left their most spectacular mark in sculpture and architecture. Temples, palaces, tombs, fortresses, bridges, roads, observatories, courts, and colonnaded halls all attest to their knowledge of engineering.

These late civilizers of the New World applied their intelligence to other things besides feats of construction. Mayan intellectual accomplishments revolved around sophisticated mathematical skills. The Maya used a number of calendar cycles, based on astronomical observations, for keeping time and as a type of divinitory almanac. Underlying it all was a strong religious motivation.

The Aztec, from their highland home in central Mexico, fashioned a different form of civilization. Theirs was an empire, primarily of political and military brilliance, with which they conquered and unified a large area of Mexico and Central America.

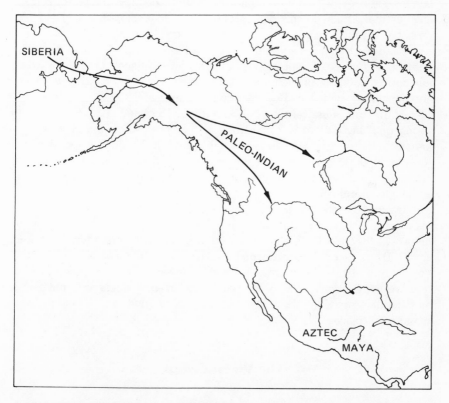

Fig. 2-2. Map of North America showing migration of American Indians from Asia and location of native high civilizations of the Aztec and the Maya.

Like the Aztec, the Inca achievement also seems to have been chiefly administrative and bureaucratic. This Andean civilization, at its height, extended from Ecuador to Chile, a dominion of more than 2000 miles from north to south. Inca power was maintained by its "state" organization with a large and well-equipped army.

In the playbill of American prehistory there are many actors. The Aztec, Maya, and Inca are only three of the most celebrated. We have given only a small sampling of the many ingredients that gave rise to those great American civilizations. Nevertheless the panorama of more than 10,000 years of American prehistory is based on solid archaeological findings.

Science responds to new evidence, particularly when it is the result of objective evaluation. Our knowledge of American prehistory is attested to by just such evidence, recovered and interpreted by high standards of scientific procedure. Yet professional scientists concerned with American prehistory are said to be embarrassed by any evidence contrary "to their already settled notions of American history . . ." (van Sertima, *They Came before Columbus,* p. 150). Mark Feldman, author of *Archaeology for Everyone* (p. 52), charges that "Even to this day, professional archaeologists and historians probably are the most narrow-minded members of academia."

Readers should recognize that these familiar protests become sterile arguments in the face of a knowledge of the history of science. One of the oldest and noblest traditions of science is its ability to be swayed by sufficiently reliable and conclusive evidence. Take this simple example:

At the turn of the century, between about 1890 and about the 1930s, a strong partiality existed in discussions of man's arrival in the New World. Leading authorities were of the opinion that the New World was populated no more than 6000 years ago. As we now know, that age has been at least doubled by the weight of slowly accumulating, convergent evidence from both archaeology and geology.

On many occasions, science has willingly discredited its own findings by insisting on precise and rigorous inquiry. The "Viking problem" is a case in point. The theory of the Viking discovery of America was first advanced about A.D. 1700 by a Swedish scholar who based his idea on the saga of Leif Ericson's voyage to Vinland. The enormous appeal of intrepid Viking voyagers has led theorists to the discovery of Viking "evidence" in such widely scattered places as Paraguay, pre-Columbian Mexico, Alabama, Minnesota, Oklahoma, and many other places.

## The Vinland Map

It would be easy to multiply the examples put forth to prove the Viking presence in America. Because most such claims are highly speculative and unsubstantiated by serious scholarship, scientists have not been persuaded. However, Vikings in America became a *cause célèbre* following an announcement by Yale University in 1965. Eight years earlier, a map of European origin had been purchased by a Connecticut antiquarian book dealer and brought to the attention of Thomas Marston, curator of medieval and renaissance literature at Yale University. The map, which was later repurchased and donated to Yale's University Library, appeared to be the only known map showing America as seen before the voyage of Columbus. The Vinland Map, as it became known, was accompanied by a notation describing the discovery of America by Leif Ericson and Bjarni Herjolfsson.

Was this astonishing map genuine? Authorities at Yale took eight years to decide. In 1965 they "went public" and gave major news services the story of what scholars called "the most exciting cartographic discovery of the century." After exhaustive testing, scholars of Yale and the British Museum, working together, concluded that the map was genuine and probably had been drawn in a Swiss scriptorium in about A.D. 1440. Yale published an exhaustive history of the investigations in a book, *The Vinland Map and the Tartar Relation* (by R. A. Skelton, T. E. Marston, and G. D. Painter), and they sold thousands of copies of the map itself. When it went on display at Yale, the map was described by a Yale librarian as "the most exciting single acquisition of the Yale Library in modern times, exceeding in significance even Yale's Gutenberg Bible and its Bay Psalm Book." Based on the many years of study and the confidence of the experts, the Vinland map truly seemed to be a welcome breakthrough in the problem of Viking visitors to America.

Then came another major discovery—a disappointing one for the academic world. The scientists had been mistaken. Although no one knew it, the evidence of a possible forgery was concealed in the map itself. A new technique for testing the age of inks had been invented. Samples of the map's ink were removed for small-particle analysis. A scanning electron instrument showed that the ink contained particles of analase, a mineral form of titanium dioxide. Electron microscopy and X-ray analysis proved that the particles were identical to commercial analase, not known before the 1920s.

The earlier conclusion, although mistaken, was thoroughly understandable. The map had been submitted to all the known crucial tests of its authenticity to coax from it all its secrets. Responsible scientists had naturally asked, how old is the ink? But the testing technique had been developed only after the map had been "authenticated."

## The Newport Tower

In a pleasant municipal park in Newport, Rhode Island, stands the Newport Tower (Fig. 2-3), a cylindrical stone structure whose origin has long been

Fig. 2-3. The Newport Tower, Rhode Island *(Stuart Scott).*

debated. Guesswork began with the opinion that the tower was the ruin of a
Viking baptistry dating from the 11th or 12th century. We have seen before
how hearsay evidence may be repeated over many years and seems to gain
certainty with its repetition. In his book, *Mysteries from Forgotten Worlds* (p.
97), Berlitz says, ". . . there is no record of who built an eleventh century
European tower on land. . . ."

Berlitz repeats the date because it is compatible with Viking exploration, and the "mystery" of the tower is allowed to live on by selectively ignoring the results of extensive archaeological excavations conducted both inside and outside the tower. Excavations were carried out during 1948 and 1949 by William Godfrey of the Peabody Museum, Harvard University. The findings were published by the Society for American Archaeology. Very careful stratigraphic control allowed the excavators to recover datable artifacts contemporary with the construction of the tower. The artifacts, consisting of pottery, clay pipes, a gun flint, and glass, called forth this conclusion from Godfrey (p. 128):

> Individually, these items may appear insignificant. Collectively, however, they are conclusive. One find, or two, might be the result of faulty excavating technique. Many finds, documented and carefully observed, excavated under circumstances which in each case preclude error, add up to a positive and unshakable conclusion. The tower could not have been built before the latter half of the 17th century. The Norse theory can no longer be entertained.

## Runic Inscriptions

Another popular approach to the Viking problem deals with stones said to be inscribed in runic writing. Runes are alphabetic characters from an early form of writing in use in Scandinavia as early as the third century A.D. Among many examples of alleged runic inscription found in North America, the best known is the Kensington Stone, found on a farm in Minnesota in 1898. Although proponents of the Viking penetration of North America still insist on its validity, the Kensington inscriptions were debunked by Danish runologists who found the writing style to be 1000 years later than its acclaimed date.

Among other artifacts from North America are elaborate hafted axeheads, once fervently proclaimed to be Norse hatchets. These have proven to be not 11th-century Viking hatchets but, rather, late 19th-century plug tobacco cutters, instruments of an advertising promotion by the American Tobacco Company (Wallace, p. 165).

## Norse Contact—Did They or Didn't They?

The reader should recognize in the preceding cases examples of how science not only discredits the oversimplified and ready-made interpretation but also, as in the case of the Vinland map, sometimes impeaches its own thoroughly researched conclusions. By continuing to suspect any hypothesis until proven, science continues an ancient and honorable practice.

On the other hand, although the specialists tend to be skeptical of new notions, they do not always refuse to accept these same new notions as worthy of investigation. It does mean, however, that any old or new idea proposed by anyone, amateur or professional, must be subject to the rigors of scientific method as we have discussed previously.

Just what is there, then, if anything, that professional specialists do accept as evidence of Norse or Viking contact? The word Viking, which means "pirate," is the term applied to those adventurous Scandinavian sea rovers who raided and conquered coastal kingdoms of Europe from the eighth to the tenth century A.D. They added another chapter to their voyaging history by expanding west to Iceland and Greenland. From there, Viking expeditions to the New World (North America) are described in the so-called Icelandic sagas, legacies of Viking history written mainly in the 13th century.

It was from these sagas that we first learned about "Vinland," a land west of Greenland. Because the sagas mention the presence there of grapes and wine, many Viking "detectives" found appropriate circumstantial proof that Vinland or Vineland, as it is sometimes spelled, must be located somewhere in the north temperate zone of warm to hot summers, perhaps as far south as Virginia. But Viking historian Helge Ingstad stressed the importance of language. In *The Quest for America* (p. 176) he cites the opinion of linguistic experts that

> The name Vinland has nothing to do with grapes or wine but that the syllable "vin" is the old Norse word for grass fields, found in so many place names in Norway, Sweden and the Shetland Islands. . . . In other words, Vinland means "The Grassy Land" and it was precisely pastures for their cattle that the Vinland voyagers were in the first place interested in finding.

In his search for proven remains of the Vikings in the New World, Ingstad, himself a Norwegian, explored North America's Atlantic coastline, always with an eye on the literary descriptions of Vinland. Finally in Newfoundland he made the discovery. On the northernmost tip of this eastern island of Canada, he found L'Anse aux Meadows, a site which to date is the only proven Viking settlement in America (Fig. 2-4).

What does the proof consist of? Artifacts of the Viking type which survived at L'Anse aux Meadows were a small spindle whorl for the spinning of wool fiber, a bronze pin, smelted copper, a stone lamp, a bone needle. These and other artifacts were found in association with a complex of multiroomed houses. Structural features excavated by Icelandic, Norwegian, and Swedish archaeologists bear a close relationship to well-known Norse remains in Greenland, Iceland, and Scandinavia. From excavated fireplaces, cooking pits, and an iron-smelting workshop, ten samples of charcoal, bone, and turf gave an average radiocarbon date of about A.D. 1000, in good agreement with the known Viking age.

Excavations, enriched by careful research, have produced much to sup-

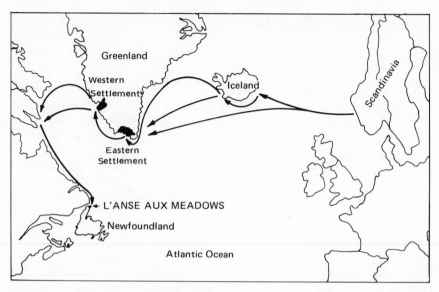

Fig. 2-4. Map of North Atlantic Viking migration routes to Iceland, Greenland, and North America.

port the general acceptance of L'Anse aux Meadows as the only genuine Viking site yet discovered in the New World.

So we know now that seaborn Europeans made their way to North America before Columbus did. The evidence, even though presently restricted to a single site, is nevertheless persuasive. But as we saw in the preceding chapter, others subscribe to more tenuous theories of early European visitors to America. The February 1977 issue of *Reader's Digest* carried an article by Thomas Fleming entitled "Who Really Discovered America?" The article ends with a firm conclusion (p. 73):

> For the first time, we must include in our American heritage fighting Celts from Spain, and daring Semitic seafarers from Carthage, Libya and Egypt.

Should we take this statement as seriously as its author does? Let us look at the evidence.

## Mystery Hill

The hill is wooded and rocky. Although out of sight, the Merrimack River flows nearby through this southeastern corner of New Hampshire. At some time in the past the natural beauty of this hilltop was intruded on by man, who came and left a complex of stone structures, the remains of a human community now known as Mystery Hill, near North Salem. Just a few miles away, along Route 28, the following statement heavy with qualification appears on a permanent roadside historical marker:

### MYSTERY HILL

*Four miles east on Route 111 is a privately owned complex of strange stone structures bearing similarities to early stonework found in Western Europe. They suggest an ancient culture may have existed here more than 2000 years ago. Sometimes called "America's Stonehenge" these intriguing chambers hold a fascinating story and could be remnants of a pre-Viking or even Phoenician civilization.*

New Hampshire State Historical Commission

The qualifiers, "suggest," "may have," and "could be," convey a great uncertainty about Mystery Hill. Because no one is certain who built the crude stonework, the vacuum has been filled with emphatic arguments favoring pre-Columbian European and Mediterranean origins. These arguments have been so strongly stated in popular publications about Mystery Hill that the more plausible viewpoint has not reached the public with the same impact.

The ruins themselves consist of a maze of 22 stone slab structures covering about one acre. In some cases the dry stone chambers are of corbeled construction, meaning a gradual projection or stepping of stones inward (Fig. 2-5), and in other cases, single large slabs or boulders form walls and ceilings. A sod covering gives some chambers the general appearance of underground caves. Outlying scattered stones and low rock walls, spread over 12 acres, are also part of the site (Fig. 2-6).

### Early Tradition and Theory

In 1826, Jonathan Pattee, a member of one of the 18th-century settler families of French Huguenot ancestry, chose the Mystery Hill site as a farmstead and built a house on the hilltop where he lived until his death in 1848. Thereafter known as Pattee's Caves, the site, with its masonry "cavern" rooms, was believed by some to have been constructed by Pattee himself. Others, including family descendants, maintain that some of the ruins had been built and abandoned before Pattee's occupancy.

In 1933, William B. Goodwin of Hartford, Connecticut, bought the land and with it the stonework puzzle of Pattee's Caves. It did not, however, remain a mystery in his mind for very long. It was his conviction that the crude stonework was the remains of an Irish monastery founded by sailor monks and dating as early as the ninth or tenth century A.D. In pursuit of his beliefs, Goodwin excavated and, unfortunately, also reconstructed the ruins to fit his

Fig. 2-5. Author Cazeau examines alleged Celtic chamber at Mystery Hill, New Hampshire *(Stuart Scott)*.

ideas. This "doctoring" of the site later led to confusion over the question of its original form. Goodwin's reconstruction was the beginning of a modern folk tradition, published by him in 1946 in *The Ruins of Great Ireland in New England.*

Six years later, in 1952, the North Salem site was given, if not more credence, at least more publicity by Frederick J. Pohl in his book, *The Lost Discovery: Uncovering the Track of the Vikings in America.* Pohl's hypothesis—like Goodwin's, more inspirational than objective—was that the ruins "may be the work of Europeans who came over before the Vikings."

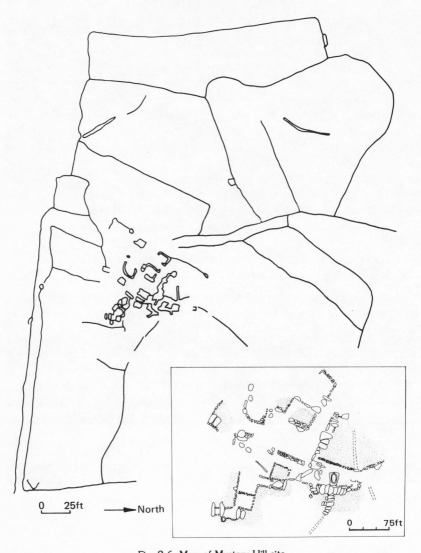

0    25ft        ➤North

0        75ft

Fig. 2-6. Map of Mystery Hill site.

In 1957 the Pattee's Caves site was bought by Robert Stone, who renamed it Mystery Hill and opened it on a commercial basis the following year. Belief in the early European origins of Mystery Hill has extended to the present time. Popular descriptions in the truth-or-fiction genre have made Mystery Hill one of the most widely publicized sites of eastern North America.

## Modern Studies

Shortly before the site was purchased by Mr. Stone, a detailed archaeological investigation was conducted at Mystery Hill by Gary Vescelius under the direction of Dr. Junius Bird of the American Museum of Natural History. Since that time a large part of the continuing study of the site has been carried out by an organization of amateur archaeologists, the New England Antiquities Research Association, or NEARA. The association was founded in 1964 out of a concern for the preservation and study not only of Mystery Hill, but of similar stone remains throughout New England.

NEARA represents a point of view, namely, that there exists in New England evidence pointing inescapably to the presence there of Old World megalith builders in pre-Viking times. NEARA states its doctrine in assertive and confirmatory terms, sometimes engaging in some mild mudslinging at the professional archaeological community. Mark Feldman, a spokesman for NEARA, devotes a chapter to that organization in his book *Archaeology for Everyone*. A relevant section reads (p. 133):

> No one can now deny that NEARA has succeeded in developing convincing archaeological, astronomical, cultural, and epigraphic evidence of European travelers in America before the time of Christ. Some professional researchers, responding to NEARA's accomplishments, have offered nothing but praise, while others still cling to the belief that American colonists were responsible for any and all unusual archaeological features. Norman Totten, professor of history at Bentley College in Waltham, Massachusetts, recently commented, "the attribution of these structures and calendar stones to colonial construction is patently absurd." Dr. George Carter, professor of geology at Texas A & M University, observed that "we cannot expect today's 'experts' to suddenly admit that they have been wrong all these years and to begin rewriting all their notes, papers, and books."

In a stronger tone, Feldman adds (p. 128):

> The Mystery Hill site stands out as an example of arrogance and authoritarianism on the part of the professional archaeological community, but also as an example of what an amateur can accomplish if he doesn't mind being laughed at.

## The Other Side of the Coin

Obviously, there are strong disagreements over the primary questions of who built Mystery Hill, when, and for what purpose? Our chief aim in this chapter is not to condemn amateur scholarship, but to encourage the average reader to look at evidence objectively and to apply whenever necessary the concepts of fact and logic discussed in the Introduction. For example, the reader should see in the final sentence of Mr. Feldman's statement another case of the fallacious argument of appeal to pity. Bergen Evans once wrote in his book *The Spoor of Spooks and Other Nonsense* (p. 32), that being laughed at doesn't make you right. A demonstration of truth will work as well in the presence of believers or skeptics.

Actually, whether or not amateurs are scorned or ridiculed by some professional specialists is beside the point. The major point of difference is that what some amateur groups such as NEARA find to be persuasive evidence of transoceanic influences, professionals do *not* find persuasive. The rather startling proposal of prehistoric European origins for Mystery Hill rests on a number of principal criteria. According to Feldman, the best evidence comes from archaeology, astronomy, and epigraphy. Let us take a closer look at these lines of evidence.

## Archaeology

Archaeological evidence consists of architecture, portable artifacts, art, and other remains of man's past cultural behavior. Architecture is the most obvious feature of archaeological import at Mystery Hill. The view that this site and others in New England are the results of a migration of Old World megalith builders is based on interpretations of similarity or dissimilarity of construction. The Mystery Hill Tour Guide pamphlet states (p. 1):

> Settlers during the pre-Colonial and Colonial periods would certainly not have expended so much effort and energy to build such sophisticated structures; indeed, the architecture is grossly dissimilar to anything Colonial. In addition to this, it was found that this site does match structures in Portugal, Malta, and Spain. Some of their structures are similar in many respects with [sic] those at Mystery Hill, hence it was deduced through comparative archaeology that megalithic (i.e., stone building) cultures must have been responsible for this site's creation.

This statement contains the assertion that people in Colonial America would not have expended the energy to build sophisticated structures. If that were true, why should we assume, as NEARA does, that more ancient people expended *their* energies for building such structures? The truth is that Colonial Americans did build structures of stone. For example, the Shakers, an offshoot sect of English Quakers, were responsible for admirable stonework. Their small utopian communities flourished from the late 18th century and one can still see examples of substantial barn, house, and shop foundations as well as bridges and waterworks in stone. The Shaker dwelling house at Canterbury, New Hampshire, has steps carved from single enormous blocks of granite. Elsewhere, dry laid granite fences, made by Shaker stonecutters, were built of stone blocks nearly a yard across. Even before the Shakers, the tradition of Colonial stone construction was established. The foundations in stone at the Fort George settlement in Maine date from 1607, and the Whipple House at Guilford, Connecticut, was built in 1629. In short, a conscious knowledge of the use of stone in Colonial America argues against the uniqueness of Mystery Hill architecture.

And here is another assumption, this one tinged with "scientific" findings. The guidebook says that because Mystery Hill bears some resemblance to structures in Portugal and Spain, deductions from comparative archaeology certify that early European megalith builders constructed the site. The evidence for this is weak and shows little knowledge of the true nature of comparative archaeology. NEARA takes selected and general resemblances for fact. We could do the same. For example, in Figure 2-7, Mystery Hill is compared with several architectural cross sections from sites of pre-Christian Ireland. Are these astounding clues to the origin of Mystery Hill? Before deciding, the reader should consider an equally plausible answer. Natives of Ireland were among the first settlers of the state. As history records it, the pioneer Colonial settlers of southern New Hampshire arrived in the early 1700s. These hardy Irish immigrants, along with some descendants of English Puritans, found familiar physical features. Pleistocene ice sheets in Ireland and in New England had left millions of scattered boulders on the land. This glacial drift material was a nuisance in the clearing of land for planting, but at the same time these boulders and slabs were an endless resource for the building of walls and structures. These people had brought with them the customs and building skills of Ireland. Undoubtedly, they also brought the memory of chambered structures and other impressive Bronze Age monuments of Ireland. Applying the principle of Occam's Razor, could we not be seeing at Mystery Hill a similar use of stone that resulted from building the dry laid construction they knew so well?

Now another immediate question arises. For what purpose were these masonry "caves" built? Vescelius, drawing on his excavation knowledge of the site, suggested that they may have served variously as food storage vaults, as foundations of frame structures, such as Jonathan Pattee's farmhouse, as animal shelters, or some related farming activity. These explanations are not accepted by those who favor a more remote origin. The megalithic walls, they say, are unique to Mystery Hill, but other researchers deny this. Birgitta Wallace, research assistant at the Carnegie Museum, Pittsburgh, and an authority on controversial prehistoric finds in America, comments on the low stone walls in *The Quest for America* (p. 173):

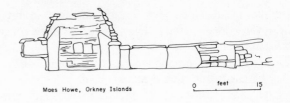

Maes Howe, Orkney Islands

0    feet    15

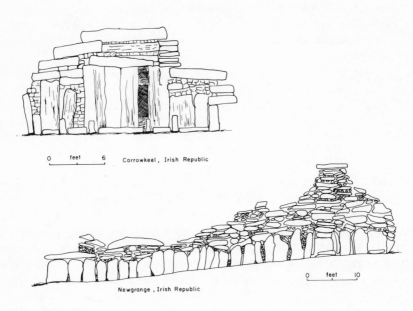

0    feet    6        Carrowkeel, Irish Republic

0    feet    10

Newgrange, Irish Republic

Mystery Hill, North Salem, N.H. (U.S.A.)

0    feet    10

Fig. 2-7. A cross-sectional drawing of Mystery Hill compared with sites in Britain.

Yet I have personally seen, throughout New England, regular stone fences built in exactly the same way as the North Salem (Mystery Hill) walls; in fact they occur wherever large boulders are particularly plentiful, which makes their occurrence at North Salem entirely logical.

Explaining the architecture of Mystery Hill in terms of European megalith builders is an unreasoned conclusion. An observed resemblance is exalted as "proof." Rather than presenting sound inferences from established fact, NEARA gives us archaeological fiction.

Now let us look briefly at another archaeological subject—portable artifacts. There is a generally accepted assumption in archaeology that artifacts will be left behind by their departed creators, the residue of their activities. Actually this is more than an assumption; it is the bedrock principle of prehistoric research, for without the material remains of the past there would be no archaeology.

Like architectural structures, which we can think of as large artifacts, smaller portable artifacts can be compared for details of design, materials, and function. We can date artifacts, then use them to date other things. Artifacts can tell us something of the people who made them. For instance, design elements on pottery have been used to demonstrate an extinct matrilineal descent system—that is, the finding of clusters of design elements within a special part of a site demonstrates the transmitting of the preferred design style from mother to daughter, as is done by present-day Indian descendants of those prehistoric pottery makers.

An artifact will often reveal, by its shape or decoration, its cultural origin. For example, an artifact is said to be *intrusive* if it appears as a new introduction from an outside source. In Figure 2-8, we see an elaborate vase found in an excavated tomb of the ancient lowland Maya civilization of Guatemala. Either the vase itself, or the exact idea for making it, was brought from a site in highland central Mexico, 700 miles away. Our confidence that the vase is intrusive is based on an accumulated knowledge of artifacts in both areas and crucial attention to observable *local* circumstances.

Fig. 2-8. A cylindrical tripod vessel from a Maya tomb, structure 34, Tikal, Guatemala *(From a photograph by Stuart Scott).*

At Mystery Hill, common sense tells us that its construction must have taken time—time enough for the accumulation of many portable artifacts. According to NEARA, the builders of the site were Celto-Iberians who derived their knowledge from Phoenician, Libyan, and Celtic civilizations. The Tour Guide states (p. 1) that ". . . pottery and stone tools unearthed match identically with European types of 2000 B.C."

Do the portable artifacts of Mystery Hill bear this out? Not in the opinion of professional investigators. The extensive excavations of Mystery Hill by Gary Vescelius produced over 7000 artifacts, none of which can be said to be of an early western European or Mediterranean origin. Instead, most of them date from the 18th and 19th centuries and consist of such things as pottery, glass, screws, nails, pipes, harness buckles, horseshoes, hooks, door hinges, coins, gun flints, bullets, metal buttons, milling equipment, and so forth. Other artifacts are assignable to native Indian cultures. Wallace writes (p. 172):

> . . . American archaeologists familiar with pottery and tool forms of the Eastern United States are positive that the North Salem pottery belongs to the Point Peninsula or Owasco cultures from 1000 B.C. to late pre-Columbian.

How is it then that Mystery Hill's defenders can steadfastly maintain that the site could not have been constructed by either Indians or colonists? The answer lies in the common practice of generalizing from exceptions. We could show, for example, that the coarse Point Peninsula or other native Indian pottery of New Hampshire shared certain techniques of decoration with pottery from Irish Passage graves and pottery from Iberia. But we could also show the same similarities with pottery complexes all over the world. A sweeping statement about "the identical matching with European types" is more a hope than a studied determination. NEARA searches narrowly for confirmation of a previous conviction. Such a search ought not to take precedence over the experience of trained observers who draw their conclusions from a basic storehouse of knowledge about American Indian technology.

## How Old Are the Mystery Hill Ruins?

Artifacts are vehicles for dating. If Celtic or Iberian artifacts had been found at Mystery Hill, they might have been compared with similar dated cultural materials in Europe to confirm an early date for Mystery Hill.

Vescelius used this method of dating by artifact association and described it in the "North Salem, N.H., Site Excavations" (p. 43). The china, glassware, and metal objects were found to resemble strongly specimens excavated at other early 19th-century sites in New England. But since no portable artifacts of an early overseas origin have been found, NEARA must bypass the possibility of artifact crossdating and they rely instead on radiocarbon or carbon-14 dating.

To explain briefly, carbon-14 is present in the atmosphere in a known proportion with other carbon, and is produced by cosmic ray bombardment. All living organisms absorb carbon and reflect the same proportion as the atmosphere. When a plant, animal, or human being dies, no more carbon is absorbed. After death, the carbon-14 begins to decay to nitrogen at a fixed and known rate. This means that older organic remains contain less carbon-14 than more modern remains. Carbon-14 is thus a kind of radioactive clock where laboratory measurements of remaining carbon-14 in organic material determine the general age of the object. This method is good for the past 50,000 years or so. In older objects there is no more detectable carbon-14 left to measure, and the object is said to be "dead."

The scientific confidence in a radiocarbon date depends on many things. There are sources of error that can produce a date that disagrees with other archaeological evidence. A sample can be contaminated with older carbon or carbon that is too modern. The contamination may occur in the collecting, handling, or pretreatment of the sample, or it may be natural, where, for example, humic acid in the soil may add very young carbon compounds. As a rule though, dating laboratories use rigorous chemical "cleaning" techniques to improve the dependability of the dating process.

The accuracy and validity of a radiocarbon date are also the archaeologist's responsibility. He must ensure that the sample was collected from an undisturbed deposit. The archaeologist should provide the laboratory technician with as much information as possible about the natural conditions of the deposit and any field treatment of the sample. Now, let us say that the laboratory issues a date. Radiocarbon dates are usually applied *indirectly* to the age of archaeological materials. In other words, a stone axe may be said to be 10,000 years old because a nearby piece of charcoal was carbon-dated at the same age. It may be a true age for the charcoal but a false age for the axe. An archaeological textbook, *An Introduction to Prehistoric Archaeology,* by Frank Hole and Robert Heizer, gives this example (p. 255):

> Wood charcoal and pottery associated together may not be of the same age, and a radiocarbon date of the charcoal could be misleading if applied to the ceramics. An instance of this occurred at the site of Cuicuilco in the Valley of Mexico, where charcoal from preceramic earth fill gave an age of 6715 ± 90 years but was associated with pottery known on other grounds to be 2700–2300 years old.

Either deliberate or accidental disturbances of soil deposits may cause artifacts to be relocated, for example by the action of burrowing animals or tree roots, and so forth. The laboratory publication of a date is only a first step. The interpretation of the date depends on the archaeologist's ability to establish correctly the time relationship between the dated sample and the artifact. In the September-October 1973 issue of *New Hampshire Echoes* magazine Joyce Zinn provides this description with a photograph from Mystery Hill (p. 36):

> The foreground excavation, in 1971, produced a radio-carbon date of 173 B.C. This dating, although not the oldest, was the second B.C. date to be obtained from the North Salem site. It was derived from charcoal found above man-worked bedrock.

This raises several important questions. Is the rock really man-worked? If so, does man's handiwork have to be earlier than the charcoal date? What is the nature of the sediments above the rock? Are they layered or stratified from occupation by man? Was the charcoal burned by human agency, perhaps from a hearth, or could it be a natural burning of roots from, say, a tree struck by lightning?

Geologists and archaeologists routinely ask themselves and each other such questions about published radiocarbon dates. As you can see, radiocarbon dating is not an easy matter of solving an age problem with a laboratory date.

Finally, there is one more thought for the critical reader. The investigators of Mystery Hill rely on radiocarbon dating to substantiate an early (2000 to 3000 B.C.) origin of the site. To our knowledge, three dates were derived, ranging from 175 B.C. to 2000 B.C. But at least six other radiocarbon samples were dated and fell within the much more recent historic period. These dates are rejected in favor of the popular proposition that the Mystery Hill ruin is more than 4000 years old. In any discussion of the age of Mystery Hill, all dates are logically relevant. Mystery Hill is a difficult and interesting problem, but its interpretation must rest on facts that represent the situation accurately and completely.

## Astroarchaeology

The possible use of ancient buildings or monuments in making astronomical sightings is a relatively new interdiscipline combining archaeology and astronomy. Closely related to this is the known ability of many prehistoric peoples to divide up the year by reference to winter and summer solstices (the longest and shortest days of the year), the equinoxes (days midway between the solstices), moon phases, and so forth. Astroarchaeology has been applied to, among other places, Stonehenge, England's famous Bronze Age megalithic monument. Astronomer Gerald Hawkins speculates that Stonehenge was an ancient "computer" capable of calender calculation and eclipse prediction.

At Mystery Hill, members of NEARA have made similar astroarchaeological observations. They see Mystery Hill as an observatory where the movements of the sun were recorded and the alignments related to various structures at the site. Let us take a brief look at the lines and drawings of the Mystery Hill astronomy. Describing the site as a "Celtic astronomical observatory," Barry Fell has drawn the "Calendar Circle" as it appears at Mystery Hill *(America B.C.,* p. 206). Another sketch we can use for analysis is an unscaled map that appears on the Mystery Hill Tour Guide pamphlet. Figure 2-9 is a composite of these two maps.

Is the claim correct that Mystery Hill is a solar observatory? How correct? We invited the comments of Ernst Both, curator of astronomy at the Buffalo Museum of Science. He kindly provided the computed azimuths as well as the basis for the general discussion that follows.

The azimuths drawn by Fell (Fig. 2-9) agree with the computed azimuths for midwinter sunset and midsummer sunset (the latter erroneously labeled by Fell "sunrise"). However, computed azimuths for both midsummer sunrise (again mislabled "sunset") and midwinter sunrise differ by nearly two degrees.

Since azimuths for sunset points are correct, it cannot be argued that the people who erected the stones made errors in observing the sunrise azimuths. It is true that around the solstices the sun moves slowly, so that it is somewhat difficult to pinpoint the exact date of both summer and winter solstices. Yet an error of nearly two degrees is sufficiently large for an actual observer to detect, since it involves nearly four apparent solar diameters.

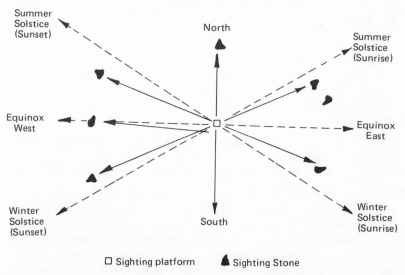

□ Sighting platform ▲ Sighting Stone

Fig. 2-9. Diagram of solar alignments at Mystery Hill, New Hampshire. Solid lines represent sighting angles as present at Mystery Hill. Dashed lines indicate angles computed by E. Both.

The discrepancies might be explained as follows: in the case of the winter sunrise the marker stone no longer stands, while in the case of the summer sunrise a broad stone is used as a marker. Several points along the rim of such a stone could be used as sights. Since Fell does not provide a scale for the apparent diameter of the sun (as is customary when dealing with astroarchaeological alignments), a definite conclusion cannot be reached.

When we turn to the Tour Guide sketch map, assuming this map is correct, it can be concluded on the basis of measurements (Fig. 2-9) that the four chief seasonal points are incorrect. Deviations from the east and west points (of the solstitial sunrise and sunset points) should be uniformly 33.5 degrees. Deviations of sunrise and sunset azimuths from the east and west points (the equinox points) as determined from the map range from 20 degrees for midwinter sunrise to 25 degrees for midsummer sunrise, while the deviations for the corresponding sunset points amount to 23 and 22 degrees, respectively. If the map is accurate, the alignments are false. However, since the azimuths given on page 206 of America B.C. are essentially correct, one is forced to conclude that the map is incorrect. In this case the entire methodology employed becomes suspect.

There is some evidence that the observing point of the alignments was not always the same. For example, in the Mystery Hill Tour Guide (p. 4), we read that the Equinox Alignment Stone "lines up in a different viewing position other than solstice lines and may have been viewed on a line parallel to the Tomb of Lost Souls." Unless we assume that this stone dates from a different period, its nonalignment with the "observing platform" would be unique among other astronomical alignments (for example, Stonehenge).

Similarly, in America B.C. we read (p. 205):

> On December 21, 1970 . . . we . . . observed the sun slowly descending towards the monolith that we had recognized as marking the solstice. From the viewing area near the sacrificial table, some 500 yards away, we saw the sun above the monolith, behind which it then set. Later researches showed . . . that the true viewing position had apparently been a stone platform 30 feet to the north. . . .

Alignments are relatively easy to find in almost any structure involving a number of points along the horizon, especially if one changes one's point of observation. Ernst Both gives this interesting example (personal communication):

> My 100-year old home is fairly accurately laid out in a north-south, east-west direction (as is very common in American houses). One kitchen window faces west; if I stand in the center of the dining room door and look toward this window, I can observe the setting equinox sun almost precisely in the center of this window. From the same "observing platform" I can observe the setting midwinter sun just grazing the western edge of a second kitchen window facing south. If I adjust my observing point suitably, I can see the midwinter sun exactly in the center of this second window. The fact

that these windows align almost perfectly with the door does not mean that the original builder was at all interested in astronomical alignments.

## Druidic Cupules

Before leaving our brief review of astronomical problems, let us take a final example from Vermont. On page 203 of *America B.C.,* Barry Fell illustrates two examples of what are purported to be "Druidic cupules." These are small circular depressions in stone, said by Fell to have been engraved by Celtic Druid priests and representing star maps. The first group is "engraved on 20-foot monolith, South Woodstock, Vermont," showing "on the left. . . part of Cassiopeia; in the center, the center polestar; upper right. . . Ursa Major."

The location of the pole star with respect to the last two stars of what Fell calls Ursa Major (actually the "Big Dipper" part of the constellation) is totally wrong. Figure 2-10 shows an accurate chart of these constellations indicating that practically all the "stars" in the Vermont example are in the wrong position.

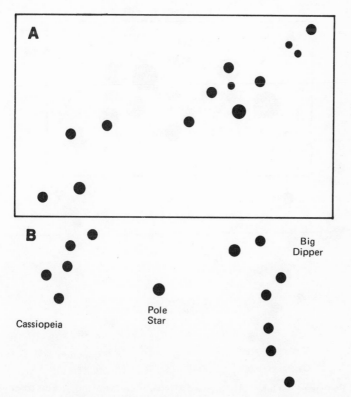

Fig. 2-10. Stone "star map" (A) at Woodstock, Vermont, compared with (B) actual appearance of constellation in the sky.

Fell also gives a drawing (p. 203) of similar cupules found at Montedor in northern Portugal. He states, "This Portuguese version is apparently a map of the same region of the sky but seen in mirror-image. . . ." In reality, the Montedor version is not a correct mirror image, as even a cursory comparison with the mirror chart will show (Fig. 2-11). Taking into account that the supposed originators of the "cupules" might have been poor artists (which would be counter to what we know of the Celts), only a wild stretch of the imagination could see in these figures a representation of some polar constellations. Both Cassiopeia and the Big Dipper have very characteristic shapes and the present pole star is positioned with respect to either of these in such a way that even children can make reasonably accurate representations.

## Epigraphy

"E Pluribus Unum," "R.I.P.," "UFO," "AT Ω," "LXXIV" —epigraphy of the 20th century. Would epigraphers of the future be able to relate these

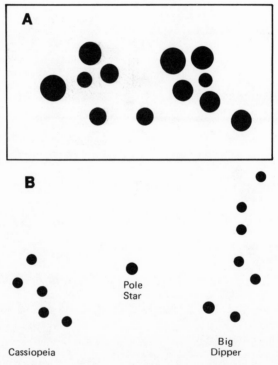

Fig. 2-11. Portugese "star map" (A) compared with (B) actual mirror image of the constellation.

mottos and symbols on unearthed fragments to the space and time framework of the cultures in which they were found? Could they say, for instance, in which century the "Blizzard of '77" occurred?

For ideas on the subject of decipherment, we invited the comment of linguist Melvin Hoffman of the State University of New York College at Buffalo. Professor Hoffman points out that epigraphy, the study of inscriptions chiseled or scratched on stone or metal, is an old and important branch of paleography, or "early handwriting." The study of inscriptions has aided the deciphering of a number of writing and symbol systems. The cracking of hieroglyphic codes contains a strongly attractive element of mystery and is a subject not easily unraveled. Hoffman reminds us of the work of the late Benjamin Lee Whorf. Whorf, whose studies of language and culture are acclaimed by modern linguists, was a long-term investigator of Mayan hieroglyphs, both incised and written. He had much to say about amateur attempts at decoding Mayan hieroglyphs. In 1940, he observed in *Language, Thought, and Reality* (pp. 196–197):

> One wrong way is to attempt a clean sweep of the job—to retire into seclusion and eventually emerge with a book—a book which tells all, which reels off, interprets, explains, epitomizes, and comments on every-thing. . . . Usually such books proclaim the discovery of a key. This key is applied at the author's sweet will, and the trick is turned on as easily as a magician lifts a rabbit out of a hat.

The recent presentation of epigraphic interpretations and conclusions reported by Barry Fell in *America B.C.* bears a disturbing resemblance to the above description. Fell declares there is epigraphic evidence that colonizers crossed the Atlantic and solemnized their presence in North America with Ogam, Punic (i.e., Phoenician), Libyan, and Egyptian inscriptions on stone.

On page 21 of *America B.C.* a tablet is presented showing resemblances between North American and European inscriptions. The tablet is said to have been found at a depth of 60 feet in a burial mound at Grave Creek, in what is now West Virginia, in 1838. According to Fell, it bears a version of the Phoenician alphabet. Despite Fell's firm belief that this proves ancient contact by sea from Europe to West Virginia, we want to bring to the reader's attention serious weaknesses in his method.

For one thing, we are given no additional information about the discovery of the tablet. except that it was found together with a skeleton and copper arm rings (neither of which points to European connections). It is perhaps not surprising that the finding of a tablet 140 years ago leaves unanswered ques-tions. For example, who was the excavator? What proof is there that the tablet could not have been forged? Where is the tablet now? Without more assurance of its validity, most archaeologists would consider the Grave Creek tablet weak circumstantial evidence.

A related point is that the single reference given by Fell for the decipherment of the American tablet is to an English epigrapher, D. Diringer. In the bibliography of *America B.C.,* we find the following listing under "miscellaneous and Curiosae" (p. 306):

> Diringer, D. 1968. The Alphabet (distinguishes the Punic, Iberian and Libyan scripts, hitherto confused by most American collectors).

Rather than being a useful bibliographic reference, this is another assertion to prop up the theory of linguistic parallels. The interested reader may want to check the Diringer source, but no publisher or place of publication appears.

Several chapters in *America B.C.* relate to the decipherment of Ogam in Ireland. The discussion is overdetailed and, it seems, out of place in a book that is represented as providing new insights into North American prehistory. Fell's reconstruction brings adventurous Celtic voyagers from the British Isles to Spain and Portugal and on to America. He has a chapter (Chapter 6, "Clues Lead to Spain and Portugal"), important to his later North American arguments, about the existence of Celtic inscriptions in Spain and Portugal. On pages 74, 78, and 80, conclusions and interpretations are sprinkled with the following qualifiers: "presumably," "seem," "apparently" (twice), "most likely," "seems," "suspect" (twice), "may represent," "only a guess." Yet Fell's data and arguments, including those qualifiers, were sufficient to convince him that "we now had *firm* evidence . . ." (italics added).

To Fell and his associates the inscriptions are part of an imported trait complex that includes megalithic construction, such as Mystery Hill, but this immediately raises time problems, as pointed out by Glyn Daniel, Disney Professor of Archaeology, University of Cambridge, England *(New York Times,* March 13, 1977):

> Professor Fell's assertion that there is Ogam writing in Spain and Portugal and that it was taken by the megalith builders from Iberia to America is not true. There are no authentic Ogam inscriptions outside the British Isles; the megalith builders of Western Europe were not Celts. They predated the Celts by over a millennium. . . .

A professor of English and folklore at the State University of New York at Binghamton, Wilhelm Nicolaisen confirms Daniel's chronology by pointing out that Ogam did not exist in Ireland before the fourth century A.D.—much too late for Fell's alleged inscriptions in America.

Objects found at Mystery Hill said by Fell to be inscribed in both Phoenician and Ogam are offered as proof that Baal, the Phoenician sun god, and Bel, the Celtic sun god, were one and the same and were worshiped on this New Hampshire hilltop. Fell asserts rather than defends his position. The following passage is typical (p. 91):

As events quickly showed, we had the solution in our hands. Within ten days we were seeing dozens of Ogam inscriptions. . . . It became clear that ancient Celts had built the New England megalithic chambers and that Phoenician mariners were welcome visitors, permitted to worship at the Celtic sanctuaries and allowed to make dedications in their own language.

One of the objects from Mystery Hill is a large boulder, designated the Beltane Stone, said to contain a date in Roman numerals, possibly carved by a visitor from Gaul (Fig. 2-12). One of the present authors (Cazeau) closely examined the Beltane Stone at Mystery Hill. It is a metamorphic rock, called a gneiss, injected by granite blebs. Following its formation this rock was subjected

Fig. 2-12. A. The Beltane Stone of Mystery Hill, said to mark the Celtic Mayday festival, in Latin inscription *(After B. Fell,* America B.C.). B. Natural strains in rock produce fracture patterns that strongly resemble Fell's Roman numerals, but they are natural *(Robert Hatcher).*

to stresses which generated what geologists call slaty cleavage, a quite ordinary geological secondary feature seen in rocks all over the world. Two such cleavage directions intersect to form three Xs which have been interpreted by Barry Fell and the owners of Mystery Hill as the Roman numeral 30. They are not. Close examination shows that they are not superficial traces, but extend through the entire rock. Other lines of cleavage form, according to its believers, Roman "I"s which are slanted to the left at an approximate angle of 30 degrees. We have never seen Roman numerals slanted; they are upright. Fell sees the Roman numerals as the number 39. It is no such thing. We might add that these cleavage directions which form the supposed numeral 39 have been emphasized by retracing with chalk, while other cleavage traces that do not fit in have been ignored.

Another example that we examined at Mystery Hill is the Eye Stone, said by Fell to be the inscription sign of the eye of the sun god Bel. The "eye" is actually a natural feature of granite injection, made somewhat pronounced by differential weathering. The shape of the granite bleb intruded along linear planes of foliation is indeed "eyelike" and is closely related to what geologists call *augen gneiss*. The word *Augen* is German for "eyes." There is no trace of human workmanship in this rock (Fig. 2-13).

Our final example of epigraphy is a slab of sandstone from Virginia, purported to be solid evidence of Iberian visitation to the New World long before Columbus or the Vikings. We examined the Virginia sandstone inscription on display at the special conference entitled "Ancient Vermont" held in October 1977 at Castleton College, Vermont. The stone depicts a scene of a sun setting behind several rounded hills, one of which is dotted with several pine trees. In the upper left-hand corner of the slab is alleged Iberian writing, translated by Barry Fell to read, "the sun sets in the notch of the hills opposite the place of worship." The inscription appears to be a fake. The naturally weathered surface is a dark brown while the inscriptions are a fresh light tan. It would require less than a century for the incisions to assume the same weathered character of the rest of the sandstone, even taking into account the rigorous weathering conditions that prevail in Virginia. Barry Fell has admitted (in a personal communication) that the inscription is of fresh appearance, but he claims that the rock was found face down in soil and was thus protected from the ravages of weathering. In actual fact, the inscription would not be protected in a face-down position. Of critical importance in all chemical weathering is the presence of water, whether the rock is exposed to the air or not. There is plenty

---

Fig. 2-13. A. Is this the eye of the Celtic god Bel, as Fell and his associates claim? The authors' examination of this stone revealed the features to be natural, rather than artificial, as can be seen in many other examples, such as B, in metamorphic rocks from South Carolina. The rocks at Mystery Hill are metamorphic *(Robert Hatcher)*.

of water in Virginia, and this rock, found near the surface, was, like all rocks in the area, repeatedly soaked and subjected to chemical weathering. Conclusion: the inscription is modern.

The reading of *America B.C.* reveals fallacies of false assumption, failure to specify, and speculative argument. Rather than try to disprove early European and Near Eastern contacts with America, we question Fell for concluding too much from too little. Benjamin Lee Whorf, cited earlier, stated a conclusion with which many professional linguists concur (p. 197):

> . . . linguistic findings must eventually bear the scrutiny of, and become the ground of, collaboration for various linguistic scholars. One man cannot be the medium for interpreting a literature; such a task requires the mutual contributions of many scholars.

## An Open-Door Policy

Mystery Hill has been presented as a case in point where the popular press has offered explanations that are seriously questioned by most archaeologists. But is it not possible, comes the eternal question, that real voyages across the Atlantic were made in pre-Viking times? Of course it is possible, maybe even probable. Notice that we do not reject the possibility. But that is not the issue. Instead, we have argued against the "excluding" philosophy of knowledge, which in the hands of some enthusiasts acquires the character of "factual truth." We suggest to the reader the need for a critical-mindedness about Mystery Hill or any other site that is similarly presented. Even though questions remain to be answered, such sites as Mystery Hill can be better understood by the application of experience and the recognition of fallacies of reasoning.

The door of Old World—New World contact is left open, but always the key archaeological questions must be asked:

First, *is the evidence of human activity truly the product of human manufacture?* In the popular press there is a fondness for the word "evidence"—something that furnishes proof. The wise reader will know that he can often substitute the word "data"—something used for the basis of discussion. Here is an interesting example.

In *America B.C.* Barry Fell devotes a chapter to ritual phallic monuments. Using photographs and drawings, he calls attention to elongated stones from North Africa, Brittany, and South Pomfret, Vermont. The only apparent feature common to all is the fact that they are each longer than they are wide. Fell boldly offers these as *evidence* of Celtic fertility cults dedicated to the power of reproduction.

He attaches special importance to a series of engravings "on the base of a carved stone phallus, four feet high, Temple of the Marriage Rite, South Woodstock (Vermont)." As *evidence* of their significance in fertility, we are told that these are diagrams representing, first, an erect phallus, followed by intromission, a swelling uterus, and, finally, the growing embryo attached by the umbilical cord to the circular placenta (Fig. 2-14). Other marks in stone from South Woodstock inspired Fell to write (p. 223):

> In the same hilltop area we found libation bowls cut into the bedrock (and partially filled by rainwater), the adjacent portion of the exposed bedrock cut into what has the semblance of a ritual copulation table, the stone excavated into a negative mould of the buttocks of the two marital partners, seated side by side. This sanctuary, for as such it evidently is to be regarded, we had at first called Phallus Hill, but after considering all the

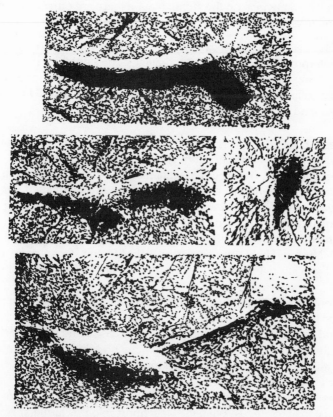

Fig. 2-14. Are the above examples artifacts proclaiming esoteric fertility rites? Or are they natural? They are quite familiar to geologists and are known as load casts brought about by uneven deposition during sedimentary rock formation *(After B. Fell,* America B.C.).

relevant facts we could elicit from the surviving relics. . . . felt that a more
appropriate name might be *Hill of the Wedding Rite,* and so we marked it
on the map, where no name had stood before.

Readers risk being badly misinformed if they accept such explanations as
*evidence.* Essential to Fell's reconstruction is the freedom to interpret shapes in
stone as art in imitation of reality. To Fell, the stone is a medium of expression
that has been used by a Celtic artist to portray a pattern of human behavior, in
this case a sexual one. Fell's research is faulty in at least two important ways:

1.  The data are not presented in a useful and comprehensible manner. Of
    the 12 pictures in his chapter, only 4 are scaled illustrations, so com-
    parisons are not possible. Furthermore, in the case of the "engrav-
    ings," four are shown, but we are not told whether these are the only
    such engravings on the stone. Maybe there are others that could not
    be interpreted as sexual symbols.

2.  By overlooking other explanations, he oversimplifies and ignores other
    interpretations. The four "engravings" described by Fell as "an un-
    mistakable sequence of necessities for the production of a child," are
    actually examples of a natural phenomenon, recognized by geologists
    as load casts or other bedding plane features.

A second key question is: *What artifacts are found, and do they reflect the
economic or other historical factors of the site and its occupants?*

In *America B.C.,* Barry Fell has written that near Davenport, Iowa, around
3000 years ago, there was a residential colony of Egyptians, Libyans, and
Celtic Iberians. Fell's proposition includes other bands of roving western
European and Mediterranean peoples who crossed the Atlantic and colonized
North America. Some came from Spain and Portugal and created such settle-
ments as Mystery Hill. Later, according to Fell, came Phoenicians, Basques,
and the Egyptian mariners by way of the Gulf of Mexico and the Mississippi
River.

Before being carried away with the idea of Egyptian boats moving up-
stream on the Mississippi, the reader should ask what the Egyptians and others
left behind. Fell's evidence is epigraphic—in the case of the Davenport Colony,
stone with engraved signs on it. But why have no Libyan, Egyptian, or other
North African portable artifacts been found?

Ancient colonists in Iowa would have shared with all fellow human beings
the necessity to cope with the physical environment. Man creates or brings a
technology to fit his needs. Were the Davenport colonists farmers? No farming
implements have been found. Neither are there subsistence artifacts such as
projectile weapons to show that they may have survived by hunting. How long
did they stay? What kind of structures protected them from the elements? Did

they come unarmed? Did they make war? Or did they perhaps find themselves colonizing in peaceful coexistence with local Plains Indian tribes? Did they intermarry? No myths or legends record Old World influences in Plains Indian populations.

Man must make provisions for himself in adapting to his surroundings. Archaeology has repeatedly shown us that even among less-developed societies such provisions are evident in the form of settlement patterns, crafts, tools, and so forth.

Why no artifacts were found is indeed an important issue. It is not a rhetorical question, but rather a question that is meant to be answered. Feldman, in a discussion of Viking travelers in Oklahoma *(Archaeology for Everyone* (p. 118), states:

> Another argument concerns the fact that no artifacts relating to Norse occupation have been found. What the professional archaeologists don't say is that no excavations have ever been undertaken!

Fair enough perhaps, if one grants Feldman's assumption that all artifacts lie buried. But much archaeological material lies uncovered on the surface. Erosion cycles have exposed many artifacts large and small, both younger and older than the Viking-age material. Feldman further quotes Mrs. Gloria Farley, an Oklahoma resident:

> Gloria Farley estimates that any Norse exploration party would probably have been limited to maybe thirty-five or so persons, and that tools and artifacts would probably not have been left behind—certainly in no quantity to speak of.

It is interesting that Mrs. Farley speaks of quantity. Quantity is a variable that doesn't seem to bother her fellow amateur, Barry Fell, who, on the basis of one engraved tablet of dubious origin, reconstructs a tableau of Iberian and Punic speakers living in Iowa in the ninth century B.C. and making use of a stone calendar regulator with Egyptian hieroglyphs. The colony arrived, it is said, by way of the Mississippi River on board ships commanded by a Libyan skipper of the Egyptian navy, during the 22nd or Libyan Dynasty. They were accompanied by an Egyptian astronomer-priest.

We have already seen that such a story as this is not at all unusual in diffusionist literature. It does not necessarily represent historical reality, no matter how firm the story is made to sound. Not only does Fell not ask key archaeological questions, but he accepts the inscribed tablet as historic fact. In this case, the tablets are known to have been forged during the 1870s, a fact fully described in "The Davenport Conspiracy" by Professor Marshall McKusick of the University of Iowa.

As to Mrs. Farley's suggestion that perhaps no artifacts were left behind, it is interesting to note historic examples of material discarded or lost by exploring parties. In the spring of 1541 the expedition of Spanish explorer Francisco de Coronado moved across the central plains from New Mexico, eventually reaching as far as the highlands of central Kansas. How do we know this? The Museum of the State Historical Society in Topeka houses a sword left behind by a member of the expedition, and examples of Spanish armor have been found elsewhere in Kansas. Coronado's expedition was not for the purpose of colonizing. Instead, it moved rapidly across those endless plains in a futile search for the mythical wealth of Gran Quivira. Even America's earliest inhabitants, the Paleo-Indians, left behind a relative abundance of artifacts, considering their rather simple lifestyle more than 10,000 years ago. To propose that North American colonies of more materially advanced peoples came from the Old World, without artifactual evidence, lies, it seems to us, in the realm of wishful thinking.

CHAPTER 3

# Ancient Astronauts
## Thesis

*Meanwhile, back on earth. . . .*

—EDWARD WELLEN

## War between the Planets

The original cause of the war is not known. It took place thousands of years ago, many light years distant from earth. The inhabitants of the earth at that time could not have known of the dramatic events transpiring in deep space that ultimately would affect them because the most advanced life consisted of simple hominids. These poor creatures were not particularly intelligent and only managed to survive precariously as hunters and gatherers, skulking by night in caves, fearful of the larger hunters that marauded in the darkness beyond the cave.

The war between the two planets raged on until a clear victor finally emerged. The victorious planet was not merciful. Learning that a spaceship of survivors from the defeated planet had fled their solar system, they sent their space legions in pursuit.

Aware of their pursuers, the survivors reached a solar system consisting of ten planets. Where to hide? Only the fifth planet was like their own, the very one to be searched first. In desperation they decided to plant a decoy on the fifth planet and hope for the best. Devices were set up quickly on the fifth planet that suggested their presence there. That done, they streaked for the third planet. It was not ideal, but they could survive there. They would have to get used to the nitrogen–oxygen atmosphere. Using thermal drills, they created caverns underground in what is now Ecuador, secreted themselves in the caverns, and waited.

63

Close on the trail came enemy space vessels. Their sensors revealed the enemy on the fifth planet. To make sure of the kill, they decided to blow up the entire planet. Powerful weapons were trained and fired on the planet between Mars and Jupiter. The planet split apart like an overripened grape and disintegrated into millions of pieces. These fragments are known by us today as the "asteroid belt."

Their mission completed, the victors departed. The astronaut refugees emerged from their caves and became acquainted with their new home, planet earth. In time, they contacted the feeble hominids and decided to help them. We do not know exactly why. By means of either artificial mutation or direct cross-breeding, a higher species sprang forth—*Homo sapiens*. Thereafter, the mutants regarded their benefactors as gods, who were portrayed in cave drawings, sculpture, and other art forms and legends. What happened to these ancient "gods" from outer space? We do not know. They may have died out or gone away. Some say they are still among us.

Such a story sounds like it came from a book on science fiction. It did not. It is a suggested scenario (somewhat embellished by us) in one of the books of Erich von Däniken. His books on the subject of ancient astronauts have sold into the millions of copies and have been the subject of television programs and numerous articles. Without doubt, there is large interest in this theme, perhaps triggered, in large measure, by our own success in space and the realization that interplanetary space travel is a reality. This opens the door to the possibility that perhaps interstellar travel is not simply fantasy. Thus, it is not surprising that the idea of contact with extraterrestrial beings, for many people, becomes something thinkable.

## The Ancient Astronaut Hypothesis

### The Basic Thesis

There have been many books written advancing the notion of outer space contact with intelligent beings. Von Däniken, who has spearheaded this movement, perhaps articulates the basic idea best. In general, his thesis is that (1) ancient astronauts have visited the earth, perhaps more than once in the earth's remote history; (2) these beings have had a profound influence on the origin and destiny of mankind; and (3) there is abundant evidence left behind of their visit(s).

The interplanetary war described at the beginning of this chapter is only one of several variations on this theme. Another possibility von Däniken offers, for example, is that this contact was not accidental, but a deliberate attempt by a superior race to lend "technical assistance" to hominids on earth, who, without their help, would have languished in a primitive state indefinitely. Still another possibility is that, like American Apollo crews on the moon, these ancient astronauts were part of an exploratory operation, with a ground crew left behind to perform experiments. Because of some snafu, they were marooned on the earth, and eventually interbred with the hominids to produce *Homo sapiens.*

## Origins of the Idea

Despite his popular identification with the ancient astronaut concept, von Däniken was by no means the first to suggest the idea. Charles Fort's writings may have been the "seed bed" from which the ancient astronauts sprouted. Fort was active in the 1920s and before. He was an avid collector of bits of information that seemed to defy all the known laws of nature and the comfortable dicta of science. Here is a typical Fortean report from one of his books, *New Lands* (1923, p. 219):

> Geneva, March 21—During a heavy snow storm in the Alps recently thousands of exotic insects resembling spiders, caterpillars, and huge ants fell on the slopes and quickly died. Local naturalists are unable to explain the phenomenon. . . .

Here is another excerpt from the same book (pp. 219–220):

> May 4, 1922—discovery, by F. Burnerd, of three long mounds in the lunar crater Archimedes. See the *English Mechanic,* 115–194, 218, 268, 278. It seems likely that these constructions had been recently built.

These cryptic words would fit harmoniously in any of von Däniken's books; however, they were written 12 years before von Däniken was born. Fort published another book in 1931 entitled *Lo!* in which there is a more specific reference to alien beings (p. 84):

> Unknown, luminous things, or beings, have often been seen, sometimes close to this earth, and sometimes high in the sky. It may be that some of them were living things that occasionally come from somewhere else in our existence, but that others were lights on the vessels of explorers, or voyagers, from somewhere else.

A later author, Brinsley Le Poer Trench, wrote a book called *The Sky People.* In the concluding words of this book, written eight years before *Chariots of the Gods?* Trench states (pp. 172–173):

> All through the ages, as you have seen from this book, the Sky People have been visiting Planet Earth and showing the way for hu-manity. . . .

Today, these extra-terrestrial visitors are to be seen in our skies in their traditional space ships. . . .

At about the same time, Pauwels and Bergier wrote an interesting book, *Morning of the Magicians,* with similar assertions. And in 1967, Robert Charroux stated *(Masters of the World,* p. 224):

Twelve thousand and five thousand years ago, astronauts came to our planet; there is no longer any doubt of that. . . .

The same year that von Däniken published his *Chariots,* the second book of Peter Kolosimo appeared, *Timeless Earth,* which addresses itself to the same theme.

It is not unusual for a long-standing idea to be revived and enjoy new popularity. We saw this in Chapter 1 in our review of diffusion—migration from continent to continent. Von Däniken's hypothesis lies at the end of a long chain of diffusionist ideas. It is, perhaps, a peculiarity of timing and the attitude of the public mind that a particular idea blossoms from obscurity and is identified with one individual.

We see the same thing happening in science. For example, in geology the concept of drifting continents was suggested first in 1620 by Francis Bacon, and later developed by Snider in 1855 and Taylor in 1908. Yet it is a theory almost exclusively associated with Alfred Wegener, a German meteorologist, who wrote a book on the subject in 1924.

## Implications of the Hypothesis

Suppose that it is true, that we are not the products of independent terrestrial evolution. Instead, we are a type of half-breed, with the blood of superior, alien beings in our veins, infused thousands of years ago, and responsible for a great uplifting of our intelligence. It is an intriguing idea with intriguing implications.

One major implication is that what man has accomplished in the past was not due to his own efforts or creativeness. Some authors visualize ancient astronauts coming out of their spacecraft and instructing primitive man directly in such matters as planting crops, domesticating animals, building cities—and even aiding in the construction of the latter.

In addition, von Däniken goes so far as to say that when we have a "brilliant" idea, it is not really our own, because it has been "programmed" into us. Thus, the paintings of da Vinci, Einstein's equations, the music of Beethoven, and so on, are manifestations of this programming. What this means is that humanity really cannot be proud of any of human achievement.

There are implications in religion, too. Is God an astronaut? Was Christ an extraterrestrial? These are questions now being bandied about. Authors make many correlations between the ancient astronaut hypothesis and statements found in holy books of world religions. The hypothesis presumes a profound effect on religions, and casts doubt on the nature of God as conceived to be by man. Von Däniken focuses on this subject in a sequel book, *Miracles of the Gods*.

We think these implications important enough, especially in view of the worldwide interest, to take a closer look at the evidence offered in support of the ancient astronaut hypothesis. Many individuals, including some scientists, quickly labeled the ideas of von Däniken and other authors as preposterous and ludicrous. Perhaps so. But many new (or revived) ideas in the history of science have been so labeled, but later proved to be valid. It is wiser, we think, to look at evidence. We will do this in the following chapter.

# CHAPTER 4

# Ancient Astronauts
## Response

*What distinguishes science from non-science is an attitude.*
—FREDERIK POHL

## Von Däniken's Challenge

Erich von Däniken, more pointedly than other authors, invites scientists to respond to his ideas. In *The Gold of the Gods,* he states (p. 84):

> I want to stimulate thought with my theories. No more, but no less. . . . I
> leave it to the scientific world to answer me.

Most of those in the scientific world have not answered him. A few have, but most of these responses are brief comments in magazines and scientific journals. A very detailed response was published in 1976: *The Space-Gods Revealed,* by Ronald Story. Ironically, Story is not a scientist. Nonetheless, his book is an admirable piece of thoughtful research. In this book, Thor Heyerdahl, at Story's invitation, comments (p. 46):

> . . . no scientist takes people like von Däniken seriously, and none of
> them cares to climb down from the academic pedestal to start discussing
> sheer nonsense merely to enlighten the man in the street.

Regrettably, what Heyerdahl says is true, and von Däniken's challenge largely falls on deaf scientific ears.

Among the few authors to take up the cudgels against von Däniken is Clifford Wilson *(Crash Go the Chariots,* 1972). Wilson makes many valid points in his analysis of von Däniken's evidence, but his treatise loses some objectivity because he has a strong religious point of view and, therefore, a special ax to grind. In some instances, his logic is as fallacious as von Däniken's, and he frequently resorts to ridicule. The question of whether ancient astronauts visited the earth in the past is not the point. The point to be made should be, is the evidence offered by von Däniken and other writers persuasive and convincing of that belief?

## Is Science So Inflexible?

A persistent theme that threads its way through the ancient astronaut literature, no matter who the author, is the apparent unwillingness of science to budge from preestablished positions. Science is even afraid, it is said, to review the astronaut evidence for fear that long-cherished ideas will come tumbling down, much to the embarrassment and chagrin of scientists who are comfortable with their present beliefs. Why this accusation? We aren't certain, but as we noted in our Introduction, these contentions often take the form of the fallacious arguments appeal to pity and argumentum ad hominen. If science can be set up as a "whipping boy," then perhaps sympathy for a cause can be aroused even though the cause has little merit.

We disagree with the premise that science is inflexible and resistant to new ideas. As we have discussed before, the scientific method demands that facts, observations, and hypotheses be rigorously tested. Conjecture and speculation are no substitutes. To assume otherwise is akin to calling a soccer team "inflexible" because they refuse to play soccer using the rules of football. Admittedly, there are pig-headed and arrogant individuals within the scientific community, but such minorities are common in most groups of people.

The history of science contains many examples showing that science responds very flexibly to the ebb and flow of evidence and new ideas. Here are a few examples:

**Carbon-14 Dating.** This method of determining the age of organic materials (such as wood, charcoal, bone, and shell) was devised in the 1950s. The public is largely unaware of the many arguments and discussions, mostly carried on in scientific journals, concerning the reliability of this method. Some dates were in error. They were corrected. Debate continues today on ways to improve this method.

**Origin of the Earth.** Early theories such as that of Buffon were accepted and later discarded. The Laplace nebular hypothesis was in vogue for a while, then discarded. The "near encounter" theories were held by many scientists, even early into the 20th century. They were discarded. Science then returned to a modified version of the Laplace nebular hypothesis. Maybe that, too, will be discarded. It depends on evidence.

**Continental Drift.** Up until several years after the Second World War, most geologists were convinced that all the continents had been fixed in their present positions since the earth's beginning. Today, most of them are sure that continents drift. Their thinking has changed also about the nature of the ocean floor. The reason? New evidence for an old theory that had been rejected.

**Pottery in the Pacific.** The discovery of some pottery fragments in a Pacific coral reef prompted a major revision of previously held notions about Pacific archaeology.

Within our own experience, the authors participated during the late 1960s in a multidisciplinary investigation of the pre-Columbian history of a region known as the Marismas Nacionales, located along the west coast of Mexico. Many scientists were involved. They included archaeologists, plant ecologists, geographers, osteologists, and geologists. What was impressive was the almost daily changes and modifications of ideas about this area as each day men and women returned from the field with new information and discussed the data with their colleagues. There were few, if any, preconceived ideas or rigid, inflexible notions.

We conclude that science *does* "change its mind." It adheres to no fixed tenets unreasonably. Science is swayed only by evidence. Therefore, it is likely that science does not regard the evidence for ancient astronauts as valid evidence.

## Evidence

The evidence offered suggesting former visits of astronauts to this planet is wide ranging. There seems to be no single piece of convincing evidence; rather, it hangs together as a tapestry of circumstantial evidence. Now, there is nothing wrong with circumstantial evidence. This type of evidence can be very strong, to paraphrase Thoreau, as when you find a fish in the milk. The question is, how strong is the evidence in the case of the ancient astronaut hypothesis?

We can divide the purported ancient astronaut evidence into the following categories: ancient stone monuments, myths and legends, drawings and other artistry, and miscellaneous unexplained phenomena.

### Ancient Stone Monuments

Did ordinary human beings build the great pyramids of Egypt? Are they responsible for the circle of large stones at Stonehenge? Were those famous statues on Easter Island carved, transported, and erected using only human labor and ingenuity (Fig. 4-1)? Von Däniken and his fellow authors claim that this could not be so. They could only have been erected with the aid and

Fig. 4-1. A. The Great Pyramid of Khufu; B. Stonehenge; C. statues on Easter Island. Some authors say humans did not, and could not, have built them (A, American Institute of Archaeology; B, Buffalo Museum of Science; C, Stuart Scott).

guidance of aliens from other worlds with advanced technology such as antigravity devices to move the huge stone blocks, and laser beams to cut them. It is further asserted that these monuments house arcane secrets that point to an "outer space connection." A usual claim is that even with modern technology these monuments could not be built today. The three stone monuments mentioned above are so well known and intriguing that we have devoted separate chapters (Part II) to discuss them. However, few of the ancient astronaut writers fail to mention other stone monuments. For example, the pyramids of the Maya in Yucatan and those of the Aztecs near Mexico City; the fortress overlooking Cuzco, the capital of the Incas, and many others.

There are certain reflections we would like to make about these stone monuments and the implications that are claimed. Astronauts came to earth. They spent much, perhaps most, of their time going about the world helping various peoples carve, move, and raise up a wide variety of stone monuments. We are at a loss for a motive. This seems a crude and laborious method by which to convey messages to future generations of mankind. Given their advanced technology, why not several plaques made of an enduring alloy that show by pictographs who they were, where they came from, why they came here, and how they came here. And also, where they were going from here. These documents could have been placed on prominent mountaintops scattered around the earth, and protected by force fields. Why come and go, leaving it to the obviously imperfect creatures here on earth (us) to later attempt to show their visit(s) by means of crude drawings and word-of-mouth tales of "gods" from the dim past? There seems to be little in the way of logic here.

We also note that these stone constructions were not all built at the same time. Did astronauts help the Egyptians build the pyramids, and then wait around for more than 4000 years to lend assistance to the Aztecs and Maya? This doesn't make much sense unless you assume that the astronauts came and went periodically. And if they did, this shows an unusual interest in our affairs spanning several millennia. In this connection, van der Veer and Moerman *(Hidden Worlds,* p. 200) observe:

> . . . what value would such a meeting have had for the extraterrestrial civilization? Let us visualize the situation: cosmonauts of a superior race from out in space, highly intelligent and skilled beings, come face to face on one of their stellar or interstellar journeys with Stone Age man who scarcely knows anything about farming, writing and so on and is not even able to communicate with his fellow men in remote places!

The astronaut writers often scoff at archaeological explanations that such and such a structure was built for religious reasons or to honor a particularly powerful or wise leader. Well, why not? This does not seem too unreasonable if we assume that early human beings were a lot like us. Today, we can see around us numerous examples of magnificent structures built for these precise reasons. There are the Washington and Lincoln memorials in America, the Vatican and St. Peter's in Rome, the Parthenon in Greece, and the Taj Mahal in India, to name only a few. No one, as far as we know, suggests that these monuments were built with the help of ancient astronauts.

On a more mundane level, think of towns you have visited. What were among the most imposing buildings there? Probably the church, or another religious ediface. Or a memorial honoring dead soldiers. We think that the reasons people erected these structures were very compelling, and compelled earlier peoples as much or more as those today.

There is one more enormous structure worth mentioning. It was not built by astronauts, it was built by men. It is a wall 1900 miles in length, sufficient to stretch from London to Moscow and beyond. It is 30 feet in height and 15 feet thick. There are 12 stone towers for every mile of wall, for a total of 24,000 stone towers. The wall extends over mountains to heights of 5000 feet, so steep in places that to walk the wall is like climbing a stepladder. From the material in this wall, we could build a smaller wall eight feet in height and three feet thick that would girdle the earth at the equator. This is the Great Wall of China.

There is no mystery about the China wall. It was built in the second century B.C., and added to in succeeding centuries. The emperor at the beginning of the construction was Chin Hwang-ti, who considered it good protection against the Huns and Tartars from the north. Impressive as the Great Pyramid of Khufu (Cheops) in Egypt is, there is enough material in the Great Wall of China to build 30 pyramids just like it.

Man's presumed incapability to build the pyramids and other ancient stone monuments must be assessed, we think, in light of such structures as the China wall and others we mentioned which we know were built by human endeavor and without help from outer space.

## Myths and Legends

A substantial part of the circumstantial evidence marshaled to support the ancient astronaut hypothesis comes from old myths and legends. Writers also draw heavily on religious literature. We regard this as a rather shaky and precarious route to truth for the following reasons:

1. Myths and legends, as it is claimed, may indeed have a basis in fact. Then again, they may be pure invention. Even if a legend was based on a real event some time in the past, nothing is easier to change by addition, deletion, or embellishment than something told and retold many times. This was especially so prior to the universality of the written word.

2. Myths and legends are subject to interpretation. Were they to be used in a strictly literal sense they would be useless as evidence of ancient astronauts. Thus, if a legend states that a "star god descended to earth riding on top of a cloud," this must be altered to read "an ancient astronaut made a landing in a space vehicle, and the ship kicked up a lot of dust as the ship landed." There is no guarantee that this interpretation is correct or even reasonable. There is always the added factor that one tends to select the interpretation that best fits one's beliefs or experience.

3. The myths and legends of the world's cultures, past and present, are legion. It is a simple matter to selectively prune from the thousands of stories only those that "prove" a point, and to ignore all others.

4. Much is made of the fact that so many myths and legends relate to skyward events. Why not? Even primitive peoples knew that things vital to their very existence occurred in the sky—heat and light from the sun, as well as rainfall—all necessary, especially to an agriculturally oriented society. Likewise, fearful things, such as hurricanes, came from the sky. Man has had many gods, and still does, that play some kind of role in his existence. It is little wonder that man has had sun gods and star gods, and gods of earthquakes and volcanoes as well. The fact that a particular ancient society paid homage to a sun god does not force one into assuming logically that this society had been visited by astronauts from elsewhere.

## Drawings and Other Artistry

Sketches of some of von Däniken's evidence of space visitation are shown in Figure 4-2. Von Däniken said in an interview with *Playboy* magazine

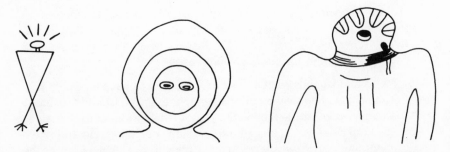

Fig. 4-2. Can these be drawings of ancient astronauts, or are there other interpretations?

(August, 1974, p. 58) that drawings such as these only became comprehensible after we had ourselves developed a space technology and the paraphernalia that goes with it. Thus, there are many drawings or other representations of creatures now recognized as being dressed up in spacesuits with helmets and antennae.

**How Strong Is This Evidence?** We do not believe this evidence to be any stronger than the myths and legends. Like myths and legends, the world contains thousands upon thousands of varied drawings, rock paintings, carvings, engravings, and sculptures to choose from which have been produced by various cultures, especially during the past 5000 years. Figurines have been unearthed that are 20,000 years old. If one wishes to do so, it is easy to assemble from this vast array of artwork a collection showing creatures who are half-man and half-lion, or half-man and half-bird. Few seriously suggest that such beings ever existed. Likewise, it is possible to assemble, as von Däniken has done, a collection of artwork that superficially resembles our own space astronauts.

But here we run into still another problem—again, that of interpretation. These drawings of supposed astronauts are being interpreted in terms of *our own technology*. We would consider it an astounding coincidence if space voyagers from the past had evolved space gadgetry almost identical to our own. Another point is that our space program is in its infancy compared with what it might be like a century from now. Would an intelligent and skilled race capable of interstellar travel possess little more in their technology than we ourselves have now?

Supposing that our space technicians had opted for square space helmets instead of globular ones. If so, then drawings such as those in Figure 4-2 would not have been produced as evidence by von Däniken. However, we are fairly confident that square-helmeted space voyagers exist somewhere in the spectrum of human artistic achievements.

As evidence, von Däniken also submits photos of two "astronauts," one holding a ray gun, it is said, and the other holding a recording disc. Once again, this is a matter of interpretation. The being holding the ray gun (this is difficult to see) also has his genitals exposed. Why would an astronaut wear helmet, suit, and gloves, but fail to protect such a sensitive part of the body? The other astronaut does hold in his right hand what resembles, crudely, a record. This is, again, an interpretation in terms of our own technology. Did ancient astronauts come across space equipped with plastic records? Our own astronauts had better communication devices in going to the moon.

Von Däniken also displays representations of a dinosaur (?) and extinct elephants. We are not sure of the point he is trying to make, except that human beings could not have been around to see dinosaurs and elephants (mammoths and mastodons). The dinosaur is a question of interpretation. It is not so well drawn that it could not also represent a lizard or a chameleon. The elephant is an elephant, but mastodons and mammoths persisted in North and South America for longer than von Däniken says. They were being hunted and killed by humans as late as 8000 B.C..

**The Piri Re'is Map.** The Piri Re'is map of 1513 is considered an important piece of ancient astronaut evidence. It is so accurate, it is said, it could only have been derived from photos taken from the air. Presumably this would be an orbiting spaceship, because the map shows South America, part of Africa, Antarctica, and the Caribbean area.

This map is not accurate. It is very inaccurate, but not bad for the time it was prepared (16th century). Others besides ourselves have looked at this map. Here is what Story has to say, in part, about the map (pp. 29–31):

> Other specific errors are the omission of about nine hundred miles of South American coastline, a duplication of the Amazon River, and the omission of the Drake Passage between Cape Horn and the Antarctic Peninsula. . . . All these discrepancies would seem to rule out the map's having been derived from an aerial photograph.

## More Evidence: The Unexplained

**Footprints in Solid Rock.** Brad Steiger (*Mysteries of Time and Space*) reports the imprint of a sandal or shoe in rock of Cambrian age (the Cambrian was between 520 and 600 million years ago), and he shows a photo of this alleged footprint in his book. What impresses Steiger is the presence of a crushed trilobite (Fig. 4-3) in the footprint, suggesting that the person who left the imprint stepped on the trilobite and crushed it.

We would comment as follows. Most sedimentary rocks, including the Cambrian rock mentioned above, were formed underwater. Trilobites, extinct now, were benthonic, or sea-bottom dwellers. There are a number of recognizable features on the bedding planes of sedimentary rock units (Fig. 4-4). The action of currents during their formation and subsequent compaction produces a variety of unusual features in sedimentary rocks, including some that might resemble shoeprints. There is nothing remarkable in the fact that the trilobite was crushed. Most fossils undergo varying degrees of deformation due to the heavy overburden of rock layers formed later. Therefore, this "sensational" discovery is a rather ordinary geological phenomenon.

**Holes in Skulls.** Von Däniken reports a bullet hole in the forehead of a Neolithic bison. We are to assume that ancient astronauts were doing some hunting 10,000 years ago with high-powered rifles (note again that the ancient astronauts use our technology). There is a picture in von Däniken's book (*The Gold of the Gods*, p. 111) of this skull, which is on display in a Leningrad

Fig. 4-3. Deformed and crushed trilobites in limestone. These are much more common than perfect specimens *(From Riccardo Levi-Setti,* Trilobites. *By permission of the University of Chicago Press).*

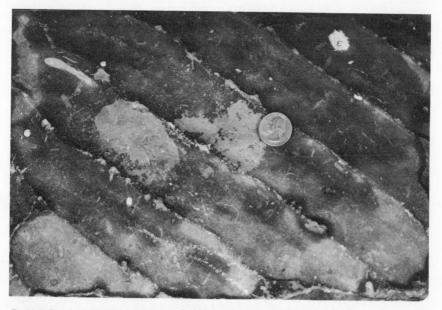

Fig. 4-4. Ripple marks preserved on sedimentary rock bedding plane. Other such features are not so obvious *(Robert Hatcher).*

museum. If this is a bullet hole, the bullet had to be more than an inch in diameter.

Holes in skulls are not exceptional. Many are natural. The dinosaurs are often classified on the basis of the number of holes (called *fossa*) in the skull. It is also well known that in some early cultures openings were made deliberately, perhaps to let out evil spirits. If we apply Occam's Razor, there are any number of plausible explanations for the occurrence of a hole in a skull, natural or artificial, without being forced to accept a 10,000-year-old high-powered rifle as the cause.

**The Nazca Plains.** One of the most striking pieces of ancient astronaut evidence, according to the believers, is the giant drawings of animals, birds, and lines ("roads") to be found on the flat desert plain of Nazca in southern Peru. These can be seen only from the air as highly stylized drawings which include a spider, condor, fish, and others. The fact that they can only be seen from the air led von Däniken and other authors to assert that they were some kind of signal or guides to incoming spaceships. Von Däniken likened the parallel lines to airport runways and parking areas.

There is no mystery as to how these drawings were made, although some authors claim they could only have been executed by supervision from the air. Maria Reiche, a scientist who has studied the Nazca Plain area, says that smaller scale models of each design were made first, and then proportioned up to full size by stretching ropes from stake to stake in a series of arcs, and removing the topsoil along these lines to expose the lighter subsoil. Reiche says some of the small-scale models are still extant next to the larger versions (quoted in Loren McIntyre, "Mystery of the Ancient Nazca Lines," p. 718).

Could such drawings have been made only with guidance from the air? We do not believe this would be absolutely necessary. Football fields and baseball diamonds are laid out without benefit of a hovering helicopter. Also, in areas where snow occurs, children often tramp out large designs of animals or other figures in newly fallen snow. These casually executed figures are often surprisingly accurate as viewed from an upper story. It is speculated that the early Peruvians could have constructed hot-air balloons from local materials and viewed their artwork from aloft, as recent experiments by the International Explorers Society suggest.

If these are not drawings that relate to ancient astronauts, then what purpose did they serve? There are a couple of possibilities. There is always the religious motive—these designs were a means of communication with gods who dwelled in the sky (but who were not astronauts). Some of the designs may represent constellations or astronomical orientations used as a calendar. Even if the artisans made these drawings as a means of self-expression, it is a more cogent hypothesis than the ancient airfield idea.

Spaceships do not need airfields. This is another example of looking at evidence in terms of our technology (modern airports). It is difficult to visualize aliens from outer space landing here in jumbo jets. Also, the Nazca Plain is littered with boulder rubble overlying soft sediment, making it unsuitable as a landing field as we understand it. A final question, asked by Clifford Wilson (*UFOs and Their Mission Impossible*, p. 84), is:

> Anyway, why 37-mile long landing strips? What was the braking system?

## How to Gather Evidence

A great deal can be learned about a scientist or other investigator by the way he or she gathers the data which forms the basis of arguments advanced. Let us compare the approach of von Däniken and that of some average anthropologists.

**Von Däniken's Methods.** Von Däniken traveled to the Caroline Islands in the Pacific to gather evidence about Nan Madol, a prominent archaeological site consisting of buildings constructed of basalt columns. The site is reached by native boat along circuitous waterways. Von Däniken complained of stifling heat and humidity so oppressive he could "hardly breathe."

This sounds familiar to us. In our experience, site work under these debilitating conditions is such that a team might spend only 4 to 6 hours out of a 14-hour day actually doing archaeology. The rest of the time is consumed by preparation, travel by boat, travel by foot, breakdowns of equipment, dealing with natives and local officials, and fighting insects, heat, and humidity. This is why extensive sites such as Nan Madol require months, even years, of investigation by teams of trained and experienced personnel.

Von Däniken arrived in the Caroline Islands for a stay of approximately one week. Taking into account that at least some of the factors we mentioned above prevailed, then von Däniken's on-site work could be measured in hours. He apparently did his work alone (no one to exchange ideas with) equipped only with a camera, notebook, and measuring tape. Yet he was able to draw the far-reaching conclusions from his data that the builders of Nan Madol had quarried and dressed the columnar basalts, and then transported them to the site with the aid of persons from another world who taught them to fly. He also proved to his own satisfaction that all previous work at the site by archaeologists was wrong, and even the carbon-14 dates could not be right.

Von Däniken's naïveté and lack of training in even the most elementary science is shown in his statement concerning the origin of the basalt columns (*The Gold of the Gods*, pp. 117–119):

> Until now scholars have claimed that these basalt slabs were formed by lava that had cooled. That seemed a lot of nonsense to me as I laboriously

verified with my measuring-tape that the lava had solidified exclusively in hexagonal or octagonal columns of roughly the same length.

Von Däniken concludes that the columns were "first class accurately worked building material." Von Däniken is apparently unaware that columnar basalts are found all over the world wherever volcanoes and lava flows occur (Fig. 4-5), and are among the common geologic features that a beginning student learns about. An error of this sort is as absurd as trying to convince a carpenter that the nails in a frame house could not have been driven with a hammer.

**A Plains Skeleton.** In contrast to the superficial, speculative, and selective approach in gathering data described above, consider what scientists did in 1974 when a human skeleton was discovered protruding from a roadway near Glendo, Wyoming. A team arrived on the site led by George W. Gill and George C. Frison of the University of Wyoming.

The soil was slowly troweled away from the skeleton with the care of a beachcomber searching for doubloons. Photographs were taken (Fig. 4-6). The skeleton was inspected meticulously. Fifteen small bones were missing. Their location was noted. The sex was determined. It had been a female with both Indian and Caucasoid traits.

Fig. 4-5. Basalt columnals. Von Däniken believes these were quarried and dressed from massive rock. They were not. They are natural, and found all over the world *(Buffalo Museum of Science).*

Six copper or brass clasps were found, and numerous glass beads. They were gathered up and counted: 11 black, 25 white, and 245 blue beads.

Fragments of grave wrappings were examined later with a microscope in the laboratory. More than 100 measurements and observations were made of

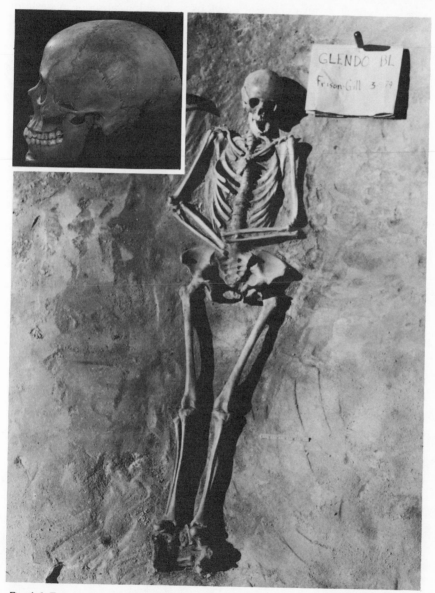

Fig. 4-6. Two views of the Glendo skeleton excavated in Wyoming in 1974 *(George Gill)*.

the skull and postcranial bones. Few details escaped scrutiny: no tartar buildup on teeth, only one cavity (the tooth was noted), the bare suggestion of an old wound.

George Gill later wrote up his results and published them in *Plains Anthropologist*. Armed with all the impressive, minute detail that had been collected as well as awareness of previous archaeological work in eastern Wyoming, Gill was in a strong position to pontificate dogmatically over the question of who this Glendo woman was. Instead, Gill presents his data and offers three separate hypotheses as to who she was, and discusses them in some detail. Where he conjectures, it is so stated. He speaks of his work as "incomplete." His conclusions are tentative: the woman was a first-generation off-spring of a white frontiersman and a Plains Indian woman. He calls for more work to be done in the Northwestern Plains region.

Gill's investigation was by no means exceptional in the way scientific evidence is gathered and evaluated. We do not see this in the techniques of von Däniken or other authors attempting to demonstrate that the earth was visited by ancient astronauts.

## Sudden Appearances and Evolution

The ancient astronaut writers point to what they consider to be a solid piece of evidence to support their theory. This evidence is the sudden appearance of man on the scene, completely equipped, it would seem, with intelligence and knowledge of agriculture, science, language, and writing. According to their thinking, this abrupt rise of humanity is borne out by the fossil record and evidence of the early civilizations. It can be explained as the result of intervention by alien astronauts who created *Homo sapiens* from a lower hominid according to one or the other of the scenarios described in Chapter 3.

In order for this thesis to hold water, the astronaut adherents feel they must reject science's theory of organic evolution, particularly as it applies to human evolution. To do otherwise is to admit that all of human intelligence and accomplishment could have been achieved by a natural unfolding rather than through the aid of benevolent aliens. This explains the rather strident attacks on evolutionary theory. For example, Peter Kolosimo (*Timeless Earth*, p. 3) states:

> Scholars of the greatest repute have upheld as a dogma what is no more than a vague and shaky hypothesis, and have constructed a whole history of human kind out of a few heaps of bones discovered in different parts of the world without being able to point to any genuine links between them. It would be easy to enumerate a host of questions and doubts concerning the bizarre theory of human origins which official science has accepted as holy writ.

We think we have shown so far in this book that science is not dogmatic, and hardly regards anything as "holy writ" unless it be the scientific method. However, the following points seem to bear discussion: evolutionary theory, sudden appearances, and the human fossil record.

## Evolutionary Theory

Organic evolution has been established as a workable theory based on a mass of evidence that few theories can match. Evolution is conceived of as an "unrolling" (as the name implies) of life on earth from simple to more and more complex forms. Thus, all life is related. The diversity we see is brought about by natural selection. Variations in offspring are brought about by sexual combination or other causes. Favorable combinations survive and persist. Unfavorable forms are weeded out and become extinct. Whether a form is favorable or unfavorable depends on the environmental conditions the organism must cope with.

Although our explanation here is greatly oversimplified, there is little argument among scientists that evolution has occurred. Many debates center on the how and why. It is the diversity of evidence for evolution that makes the theory so viable. Let us consider very briefly some of the lines of evidence that support it:

1. The *fossil record* from geology shows progression (generally) from simple to more and more complex and diverse forms with time. This evidence rests upon literally hundreds of thousands of facts.

2. *Vestigial structures* are those found in organisms that no longer have any function, but once did. These structures are not rare aberrations, but are numerous. Examples are the vestigial tail bones in man and the hip girdles in snakes (snakes once had legs).

3. In *medicine,* we have the evidence of serology. Blood is more similar, for example, between humans and pigs (both are mammals) than between humans and chickens.

4. Our own *breeding experiments* in which, instead of "natural" selection, the breeders control the sexual combinations and determine which traits will survive and become enhanced. This has led to various new breeds of dogs, cats, cattle, horses, and many other life forms.

5. In *embryology,* we see that prior to birth many organisms pass through developmental stages that are structurally similar to more primitive creatures. Human embryos have gill slits, suggesting fish ancestry.

6. *Homologous structures* indicate evolution. The bone structure of, say, the forelimb of such diverse creatures as the bat, the whale, and the mole are the same. They are all mammals. They have evolved from a common mammalian ancestor and adapted themselves to distinct environments where the forelimb has been modified to perform different functions.

Where are the famous "missing links" that so many say do not exist? They are there. We have excellent fossil remains, for example, that are intermediate between fish and amphibian, between amphibian and reptile, between reptile and bird (Fig. 4-7), and between reptile and mammal.

The evolution evidence is good—not vague and shaky as Kolosimo maintains. He and other such authors may not be aware of the evidence we have mentioned above. In this connection, van der Veer and Moerman *(Hidden Worlds,* p. 7) observe:

Fig. 4-7. *Archaeopteryx,* the intermediate "missing link" form between bird and reptile *(Smithsonian Institution).*

The fact that many people do not take these proofs seriously is not something for which science is to blame. It often happens that those with least knowledge of the subject oppose the theory of evolution most strongly. . . .

## Sudden Appearances

The primary document for life's evolution on this planet, especially throughout the past 600 million years, comes from the geologic record—thousands of feet of layered sediments, now hardened to rock, that captured and preserved the remains of creatures that lived and died at the time the sediment was accumulating. This record has been studied methodically and meticulously by scientists in every country for nearly three centuries.

There are few, if any, "sudden" appearances of life forms. These appearances are only *apparently* sudden. This is easily understood if one understands how sediments accumulate. The orderly deposition of layer upon layer of sediment can be interrupted by either nondeposition or the erosive destruction of layers (and their contained fossils) already formed. Even in cases where these interruptive processes are not operative, the change from one fossil type to another in a succeeding layer may represent a time interval of several million years. This is hardly cataclysmic.

If we are talking about sudden appearances in other contexts, the same ideas apply. L. Sprague de Camp *(Lost Continents)* puts it this way:

> Such wrong ideas about the sudden appearance of cultures arise because the relics of the later stages of a past culture tend to be more numerous than those of its early stages . . . and as these later things have not been lying around so long, the more recent relics are less likely to have been destroyed by fire, rust, termites, or vandals. For instance, if you tried to reconstruct the history of iron armor from the exhibits in the Metropolitan Museum of Art in New York alone, you might conclude that iron armor came into use suddenly in the fourteenth century in a highly developed state. But as we know from art and history, iron armor was made many centuries before 1300; the earlier pieces, however, have almost all rusted away to nothing or were turned in to the smiths for scrap.

## The Human Fossil Record

The human fossil record is not as complete and extensive as many other creatures that have lived upon the earth. There are a number of excellent reasons for this other than invoking the unworkability of organic evolution or the machinations of ancient astronauts.

Humanoids have been on the earth for little more than 2 million years. This is really not much time in which to establish a satisfactory fossil record. Compare: the dinosaurs held sway for an entire geologic era (as much as 150 million years). Yet, the finding of a complete dinosaur skeleton today is tantamount to finding buried treasure. Try to buy one.

Related to this problem is the question of fossil preservation of land creatures. Dinosaurs and man were land dwellers. Their chances of preservation were immeasurably less than, say, sea-bottom organisms whose remains are in abundance. Remains of land dwellers are quickly attacked and destroyed by scavengers and chemical decay. Note, for example, the great herds of buffalo that darkened the plains of the United States interior just a century ago. Remains of bones today are very scarce. They *did* exist. They have been mostly destroyed. So, too, with the bones of man.

It should be remembered that humans, in those early days, were few in number (unlike the buffalo). This means a smaller population of individuals available for fossilization. Also, they were wary, being perceptive of the common traps like quicksand and tar pits that fossilized less fortunate land animals.

Despite these factors that diminish the chances of preservation, the remains that have been found show an orderly development from prehuman anthropoids to *Homo sapiens.* These developments can be traced through such changes as teeth, cranial capacity, jawbone, brow ridge, other skull characteristics, associated tools, and other implements. Scientists are the first to claim that their evidence is incomplete. It is perhaps the *lack* of a dogmatic stance that invites attack by anti-evolutionists. Van der Veer and Moerman comment *(Hidden Worlds,* p. 7):

> . . . (many people) are willing enough to acknowledge that evolution applies to animals, but not to human beings. They want to exempt man entirely from its workings, and this we consider somewhat illogical. . . . If people believe that man has always existed in his present form, then it seems logical to ask where we might find the fossil remains of these people; yet no such remains exist!

We do not think, in summary, that so-called sudden appearances in the fossil record and rejection of the weight of evidence amassed in favor of evolution constitute a valid basis for advocacy of the "vague and shaky" hypothesis that ancient astronauts visited the earth thousands of years ago and helped to invent mankind and raise great stone monuments.

The formulation of a preconceived notion, the selective and biased collection of information, and the application of preferred interpretation to the exclusion of other alternative explanations does not constitute scientific inquiry and the road to truth as we understand it.

**CHAPTER 5**

# The UFO Phenomenon

*When the long-awaited solution to the UFO problem comes, I
believe that it will prove to be not merely the next small step in
the march of science but a mighty and totally unexpected
quantum jump.*

—J. ALLEN HYNEK

## The Problem of Flying Objects

Before the year is out, the Government—perhaps the President—is
expected to make what are described as "unsettling disclosures" about
UFO's—unidentified flying objects. Such revelations, based on information from the CIA, would be a reversal of official policy that in the past has
downgraded UFO incidents.

The above item appeared in the April 18, 1977 "Washington Whispers"
column of the prestigious *U.S. News and World Report*.

Few mysteries of our planet are as elusive as the UFO. Unknown objects in
the sky, by day and by night, have been seen by farmers, truck drivers, airline
pilots, lawyers, doctors, and orbiting astronauts, among many others. As the
above quote shows, UFOs have been the target of government investigation.
They have been the subject of thousands of magazine articles and books.
Newspapers around the world report sightings of UFOs almost routinely, and
have done so for many years.

What are these things? And what are they not? The UFO literature ranges
from works of sober analysis by well-known scientists to sensationalist speculation. Unfortunately, the twaddle that has been written about UFOs makes the
job of getting at the truth as difficult as trying to eat Jello with a pair of tweezers.
In addition, UFOs cannot be conjured up at will. We have to rely mostly on

89

reports after the fact, often uncorroborated, by observers of diverse backgrounds who may or may not be reliable witnesses, regardless of their standing in the community.

One can look at a photo of a figurine or painting that von Däniken says is a representation of an ancient astronaut. This is purported evidence that one can chew on. The figurine can be studied. It can be related to the culture that produced it. Historians and archaeologists can be consulted. Some fairly definite conclusions can be drawn. In contrast, a person reports and describes a UFO. He or she may be the only one who saw it. This kind of evidence is much more difficult to evaluate.

## Background

People *do* see objects in the sky. They have been doing so for centuries. The problem of interpretation again crops up (as we saw in the last chapter). What is it observers *thought* they saw, and what was the object in reality? This applies to photographic defects as well (Fig. 5-1A). Undoubtedly, the truth of many reports will never be known with certainty.

Before manned flight, objects were found in the sky, and hence they could not have been aircraft or weather balloons. J. H. Brennan *(The Ultimate Elsewhere,* p. 76) reports:

> They were seen over Florence on December 9, 1731; over Lisbon in October, 1755; over Switzerland in 1761; over France in 1779. They appeared over Pinerolo in 1808; over the English Channel in 1821; over Germany ten years later; over China in 1845. In October, 1877, eight of them flew over Wales, in formation. . . .

Some authors speculate that UFO sightings go back to biblical times. A frequent example is the description of an apparent UFO in Chapters 1 and 10 of Ezekiel. Ezekiel states (1:4):

> As I looked, behold, a stormy wind came out of the north and a great cloud, with brightness around about it, and fire flashing forth continually, and in the midst of the fire, as it were gleaming bronze.

In more modern times, the term "flying saucer" came into use, and has become almost synonymous in the minds of many, with UFO. The distinction can be made that a UFO is any unidentified object seen or thought to have been seen, no matter the cause. The term "flying saucer" implies that a particular shape—circular, disc-shaped, oval—was observed. The saucer designation began in 1947. In that year, on June 24, a civilian pilot named Ken Arnold encountered a formation of flying objects in the vicinity of Mount Rainier, Washington. Arnold, in attempting to describe the maneuvers of the

Fig. 5-1. Are these really flying objects from outer space? Could A be a photographic defect in developing? Could B be an ingenious fake? *(U.S. Air Force)*.

nine disclike objects he saw, stated that they were "flying as a saucer would if
you skipped it across water. . . ." The catchy "flying saucer" label was quickly
picked up by newsmen and has become part of the UFO vocabulary. Incidentally, Arnold's classic encounter is still listed by several authors as among the
unexplained sightings.

Even before the space programs of the United States and the Soviet Union
demonstrated that travel beyond earth was feasible, popular authors discussed
in detail the possibility that UFOs were spaceships from another planet, probably outside of the solar system. This is known as the ETH, or extraterrestrial
hypothesis. We will have more to say about this later in the chapter. The ETH
idea has become so ingrained in the public mind that many equate "flying
saucers" and UFOs with alien spaceships visiting our planet. It is easy to see the
connection between the ancient astronaut concept and UFOs. And the writings
of people like von Däniken have probably encouraged the spread of this idea.

In the face of mounting numbers of UFO sightings, the United States Air
Force became concerned and set up what was to be called Project Blue Book to
investigate these sightings. The major question they sought to answer was if
these objects in any way posed a threat to the security of the United States.

## Project Blue Book

The Air Force's Project Blue Book lasted nearly 22 years, from September
1947 to December 1969. During this span of time 12,618 reported sightings
were investigated. At the outset the program was called Project Sign. This
designation was changed on February 11, 1949, to Project Grudge. The name
was changed again in 1951 to Project Blue Book.

At the termination of Blue Book, the Air Force announced its conclusions.
The sightings, they said, were explainable as natural phenomena, hoaxes,
hallucinations, and so on. They were not of extraterrestrial origin, and they
represented no threat to the security of the United States. The files of Project
Blue Book were turned over to the National Archives, and have been available
for public examination since July 14, 1976, at Maxwell Field, Alabama.

Apparently, during the life of Project Blue Book there were some personnel associated with it who took the investigation very seriously and others who
did not. J. Allen Hynek, who was one of the scientific advisers for the program,
notes in his book *The UFO Experience* (p. 2) that even in its early stages "the
Pentagon began to treat the subject with subtle ridicule." This circumstance
may well have given rise to uneven and sometimes contradictory treatment of
data by Blue Book personnel. On the one hand, Air Force memos allegedly
stated that the UFO is "serious business" while some Air Force investigators

were dismissing well-documented, multiwitnessed UFO events as "hoaxes" or "a sighting of the planet Venus."

It was not long before charges of a "cover-up" by the Air Force emerged from many quarters. A leading voice in the claim that UFOs are alien spaceships is Donald E. Keyhoe, a former Marine major, who asserted *(Aliens From Space,* vii):

> . . . AF Headquarters, following a high level policy, still publicly denies that UFOs exist, convinced this is best for the country. But for years the Air Force has had full proof of UFO reality.

And another author, Gabriel Green, observes in his book *Let's Face the Facts About Flying Saucers* (p. 105):

> It would appear to any intelligent person that the Air Force's Project Blue Book is merely a red herring cover-up. . . . Undoubtedly the government knows more than they have released on UFO's.

Scientists got into the act, but stopped short of calling the whole thing some kind of cover-up. An outspoken critic, James E. McDonald, for example, considered the Air Force's effort totally inadequate in trying to get at the truth. He believed the effort a "grand foul-up" rather than a "grand cover-up" in his report to a 1969 symposium on UFOs sponsored by the American Association for the Advancement of Science (AAAS) (Sagan and Page, *UFO's—A Scientific Debate,* p. 54). Physicist Philip Morrison, in the same symposium (Sagan and Page, p. 288), noted with regard to the handling of UFO data that

> If the Air Force presented data in this way in a court of law, sharp attorneys would make monkeys of them. I don't believe that there is the slightest evidence of Air Force suppression of data in this whole affair, but I can well understand persons who believe there is.

## Natural Explanations

**Those with Optical Reality.** Of the 12,618 cases investigated, the U.S. Air Force came up with satisfactory explanations for 11,917 of them. Whether or not these "solved" cases are satisfactory to all is a matter of debate on a case-by-case basis. For example, the following excerpt from *Project Blue Book* (p. 156) edited by Brad Steiger rehashes a famous 1948 incident that is many times recounted in the UFO literature:

> A pilot and copilot were flying a DC-3 at 0340 hours on July 24, 1948, when they saw an object coming toward them. It passed to the right and slightly above them, at which time it went into a steep climb and was lost from sight in some clouds. Duration of the observation was about 10

seconds. . . . The object seemed powered by rocket or jet motors shoot-
ing a trail of fire some 50 feet to the rear of the object. The object had no
wings or other protrusion and had two rows of lighted windows.

It so happens that there were conspicuous meteor showers that night and
this incident was considered to be a distant sighting of a meteor according to
scientists Donald Menzel and J. Allen Hynek. This was one of the few times that
opponents Menzel and Hynek agreed about anything concerning UFOs. Of
course, editor Steiger and others are free to interpret the case otherwise. We
have no opinion about the two rows of lighted windows except to observe that
the eye does play tricks. Note, for example, that many astronomers during the
past century thought they saw "canals" on Mars. Mariner space photos
showed that the canals do not exist. It is well known that the eye tends to
connect dots into lines and irregular objects into something familiar.

Few would disagree, we think, that many sightings are due to natural
causes. Table 5-1 is a partial list (in no particular order) of the natural possibilities that might be misinterpreted by honest observers. It may appear
ridiculous to some that a rational person with fairly good eyesight could mistake
a star, or an ordinary airplane, for something else. However, we would caution
that much depends upon the angle of view, prevailing weather conditions,
relation to other objects, and so on. For example, to a nocturnal observer, a
stationary star may appear to be moving simply because it is being viewed
against a sky of fast-moving clouds.

The high degree of authenticity espoused for some sightings is based upon
the reliability of expert witnesses. In general, we have little quarrel with this
idea. We would prefer to accept the sighting of a mysterious flying object
attested to by a young and sober policeman rather than a professional liar. Yet
if the police officer, himself, has never before seen a mirage, and what he saw
may have been a mirage, he is less able to dismiss his sighting as a mirage.

Could an astronomer, of all people, mistake a star for a UFO? Yes, says

### Table 5-1. Possible Causes for Mistaken Identification of UFOs

| | |
|---|---|
| Planets | Birds |
| Bright stars | Reflections |
| Weather balloons | Artificial satellites |
| Airplanes | Meteors |
| Helicopters | Searchlight beams on clouds |
| Rocket launchings | Mirages |
| Contrails | Unusual cloud formations |
| Lightning | Northern lights |
| Kites | Parachutes |
| Insect swarms | St. Elmo's fire |

Donald Menzel, who is an astronomer. He relates the following personal experience (Sagan and Page, p. 133):

> Flying in the Arctic zone near Bering Strait on March 3, 1955, I observed a bright UFO shoot in toward the aircraft from the southwestern horizon. Flashing green and red lights, it came to a skidding stop about 300 feet, as nearly as I could judge, from the aircraft. Its apparent diameter was about one-third that of the full moon. It executed evasive action, disappearing over the horizon and then returning until I suddenly recognized it as an out-of-focus image of the bright star Sirius. The sudden appearance was due to the presence of a distant mountain that momentarily cut off light from the star. . . . I ask, how many astronomers have seen a bright star just on their optical horizon in the clear Arctic atmosphere from an altitude of 20,000 feet?

**Other Natural Explanations.** In the absence of any optical manifestation, some supposed sightings may be due to hallucination, defective eyesight, intoxication, and the power of suggestion. While we do not think these constitute a major source for reports, it is noteworthy that sightings seem to come in waves. In 1947, there were 122 sightings in the United States, according to Blue Book files. In 1952, there were 1501 reports. This was the same year in which a popular flying saucer movie, *The Day the Earth Stood Still,* was being exhibited throughout the United States. People, hearing and reading about recent reports of UFOs, especially in their area, might perhaps be more prone to be looking for something that really isn't there. Hunters, for example, on the first day of the season, are more prone to "see" deer or other prey simply because of anticipation. This is especially true of the novice hunter. Alternatively, people may be more willing to report a sighting if there is less chance of ridicule.

Under the heading of natural explanations we can list also hoaxes or other deliberate attempts to deceive. Many photographs of alleged UFOs are probably fakes (Fig. 5-1B).

## The Unexplained

At the conclusion of Project Blue Book, the Air Force still had 701 UFO cases unexplained. If more information had been available, some of these cases no doubt could have been shifted to the "explained" category. However, there is a residue of sightings, how many we do not know, that may never be explained in rational terms. There are also many cases beyond the 12,618 in the Blue Book files that were never considered because the reports did not come through official U.S. Air Force channels.

J. Allen Hynek, scientific consultant to the Air Force, claims *(The UFO Experience,* p. 144) that some cases should not have been placed in the "solved" category. One such case was the sighting of a UFO on the ground by Socorro, New Mexico, policeman Lonnie Zamora in the late afternoon of April 24, 1964.

To summarize this incident, Zamora spotted an egg-shaped object in a gulley one mile south of Socorro off U.S. Highway 85. His attention was first attracted by the sound of an explosion. From a distance of 1000 feet or more, he noted that the object was supported off the ground by four extended legs, and two figures dressed in white were present. Zamora drove to within 150 feet of the object whereupon it rose into the air amid flames and smoke and moved away rapidly to the west. Zamora was able to see some kind of insignia on the hull of the craft. Ground indentations and burn marks were found later at the site.

Dr. Hynek later interviewed Zamora and otherwise made a fairly detailed study of this case. Hynek could come to no specific conclusions except that he felt sure no hoax had been perpetrated. He considered the Air Force's official explanation—a ground light—as being a case of "creative evaluation."

Another interesting case, but not one of particular interest to the U.S. Air Force, was the reported sighting of UFOs over Papua in eastern New Guinea by an Anglican priest, Rev. William B. Gill, and several other witnesses associated with the mission at Boianai (Sagan and Page, pp. 146ff). This case is difficult to summarize because several sightings took place during the summer of 1959, and the Reverend Gill had a penchant for taking detailed notes of the events.

In essence, Gill reported seeing one or more bright UFOs with discernible "decks" hovering in the sky shortly after sunset. One of these objects was as low as 450 feet. The incredible part of Gill's report is that humanoid figures were seen working around on the decks of the UFO, and when Gill waved at them, they waved back. Native Papuans averred that they had seen these UFOs too, and signed documents to that effect.

This case generated opposing viewpoints between scientists Menzel and Hynek. No one, it seems, would argue that the Anglican priest was a hoaxer. But Menzel said that Gill was actually looking at Venus; Hynek said Gill knew where Venus was, and it was not Venus. Menzel said Gill needed glasses or was not wearing glasses; Hynek said Gill was wearing glasses. Menzel said the other witnesses were not credible because they would say they saw something just to please the Reverend Gill; Hynek insisted the other witnesses were credible.

The present authors, of course, had no direct involvement in this case, but we do have a couple of comments. Menzel speculates that the Reverend Gill was myopic and was squinting at the object (Venus). In the process, Gill mistook his own eyelashes for human figures moving around and waving. This seems to us a rather weak explanation. Most people, as far as we know, are aware that they can see their own eyelashes at times and recognize them for what they are. However, we agree with Menzel that for the Reverend Gill to look at his watch, note that it was dinnertime, and go inside and eat while these objects were still hovering and their humanoid occupants still visible, is an incredible and inexplicable action.

We can come to no conclusions at this point except to say that (1) objects have been seen in the sky and sometimes on radar, (2) some of these sightings have natural explanations, and (3) some cases do not conform to natural phenomena according to the data presented.

## The Extraterrestrial Hypothesis

### The Basic Theme

The extraterrestrial hypothesis (ETH) holds that flying saucers are space-craft controlled by superior intelligences from another planet, probably outside our solar system. A lengthy list of books could be compiled wherein this is the central theme.

It is a long step from the proposition that unidentified objects can be seen in the sky to the conclusion that they are piloted by extraterrestrial beings. Even if one accepts the idea that life elsewhere in the universe exists, which many scientists do, there should be some evidence to establish that such life is sufficiently advanced to be capable of interstellar travel, be capable of identifying and reaching our world, and to have some interest or purpose in so doing. The question might also be raised as to why these "aliens" have not established formal contact with heads of government. We merely speculate here, but it seems odd to us that these intelligences would come into our system only to remain out of reach, so to speak, but flirt coyly with just a few selected individuals.

Let us now look at some of the evidence offered by one of the leading proponents of the ETH, Major Donald E. Keyhoe, whose works are regarded almost as Holy Writ among his adherents.

### Major Keyhoe's Aliens

Major Keyhoe's books have been read widely. To his allies, his evidence is very convincing that UFOs are alien spaceships. His evidence consists of a series of case histories of UFO sightings and contacts. We selected a few of Keyhoe's reports at random to see if they were corroborated by other sources. In *Aliens from Space* Keyhoe talks about the use of "heat-force" by UFOs. On page 22, he describes an incident in which this heat-force was used, with fatal consequences as it turned out:

> The most outstanding case involved an AF interceptor crew. Just before noon, on July 1, 1954, an unknown flying object was tracked over New York state by Griffiss AFB radar. An F-94 Starfire jet was scrambled and the pilot climbed steeply toward the target, guided by his radar observer. When a gleaming disc-shaped machine became visible he started to close in. Abruptly, a furnacelike heat filled both cockpits. Gasping for breath, the pilot jettisoned the canopy. Through a blur of heat waves he saw the radar observer bail out. Stunned, without even thinking, he ejected himself from the plane. . . . The F-94, screaming down into Walesville, N.Y., smashed through a building and burst into flames. . . . Four died in the holocaust. . . .

This account seems to diverge from the *New York Times* account of this incident (July 3, 1954). According to the *Times,* two F-94s were already in the air on a practice scramble, and were asked by Griffiss Air Force Base to check out an unidentified *plane*:

> . . . the pilots had been satisfied that the plane had been "friendly" and were headed back to the base when fire broke out in the cockpit of one of the jets.

The plane did crash and four people were killed, as Keyhoe reports, but there is no mention of a confrontation between the jet and a "gleaming disc-shaped machine" armed with a heat ray.

On page 91 of the same book, Keyhoe implies that astronauts saw alien spacecraft:

> On through June, pressure on the Air Force continued. On June 4, Astronaut James McDivitt photographed an unknown device from his Gemini IV spacecraft.

but the *New York Times* on June 5, 1965 reported that "Major James A. McDivitt sighted and photographed a satellite in space yesterday." The satellite was thought to be Pegasus 2, a satellite placed into orbit by the United States to measure micrometeorite population. Incidentally, when astronauts speak of a UFO, they mean exactly that—an *unidentified* flying object, and not a flying saucer.

The most flamboyant of the UFO stories are those in which contact is supposedly made with the aliens themselves. Major Keyhoe describes the following encounter *(Aliens from Space,* p. 103):

> That same night [August 1, 1966], police investigated the story of a frightening space being, reported by a Jamestown, N.Y., girl. According to

the police, the girl was one of a group of young people in a picnic area on the Erie Peninsula. They were about to leave when an unfamiliar flying object appeared, apparently landing not far away. While the rest waited to see if it would reappear, the girl went back to her car. A few minutes later, she told police, a strange, hairy creature tried to break into the car, then climbed onto the roof. When she frantically blew the horn to bring her friends the monster jumped to the ground and disappeared.

The *Buffalo Evening News* on August 1, 1966, carried a somewhat different version of this story:

> She [Betty Jean Klem] also reported that . . . a "formless" creature appeared from the bushes and stood within five feet of the automobile. She said it was about six feet tall, had a head and shoulders but she saw no legs.

So apparently it was not a "hairy" creature but a "formless" creature. At no time did Miss Klem say it jumped on the roof or tried to break into the car. Just where did Keyhoe get this information?

The following Indiana account in Keyhoe's *Flying Saucers from Outer Space* (p. 61) was reported by the *Indianapolis Star* as a meteor:

> . . . As startled citizens stared upwards, a huge, oval-shaped machine raced out of the south-east and over the city. Barely 5,000 feet high, it was seen by thousands of people as it streaked overhead, trailed by a fiery exhaust.
>
> In two minutes police, airport, and newspaper switchboards were swamped with calls from frightened citizens. . . .

Keyhoe wrote these words in 1954. The same incident was described by Clifford Wilson 20 years later in his book *UFOs and Their Mission Impossible*. It is of interest how these writers borrow from one another, using virtually the same wording (p. 15):

> A huge oval-shaped machine was seen racing over the city, barely 5,000 feet above the ground. It was witnessed by thousands of people, and almost immediately the switchboards of the police, the newspaper offices, and the airport were swamped with citizens—many of them frightened. . . .

The foregoing excerpts from Keyhoe's writings seem to be embellishments of the original stories, if the newspaper accounts can be believed. If not, then Major Keyhoe should be more specific as to his sources. The same applies to Clifford Wilson, who apparently would rather reword the writings of the major than dig into the original accounts themselves. We think such evidence as Keyhoe presents is weak indeed.

## Captain Cathie

Captain Bruce Cathie is a New Zealander who has come up with an interesting theory which he describes in two books, Harmonic 33 and Harmonic 695 (co-authored with Peter Temm). Basically, Cathie maintains that ancient astronauts constructed a worldwide grid system buried under the ground and the seabed which was used by interstellar vessels to draw on for power and/or navigation. In modern times, the extraterrestrials are repairing and perhaps extending this grid system, or building a new grid. At least Captain Cathie *sounds* more scientific than Major Keyhoe.

Cathie sees mathematical connections between the locations of UFO sightings and the grid system he has depicted. The grid system is a series of intersecting lines strongly resembling latitude and longitude lines one sees on a globe of the earth. Cathie's books are peppered with numerous calculations of minutes of arc, radians, and harmonics. It appears to be a very scientific approach. Yet as one reads Harmonic 33, Cathie's enthusiasm for his theory rises progressively, until he is using it to explain almost everything. He locates major earthquakes and volcanoes on his grid system, and ascribes these natural phenomena to defects in the ancient grid system.

It is difficult to accept Cathie's contention that atomic bombs cannot explode unless the sun and planets are in a certain juxtaposition, which he relates to his grid system (p. 74):

> The bomb is now set up on a geometric position on the Earth's surface, and now they have to make it go bang. The position they have picked must have a certain geometrical relationship to the solar system . . . [the] answer . . . is to set up the bomb and arm it to go off when the Sun passes through the required geometric point. . . . We have cracked a code that the atom bomb countries have long tried to hide.

We are not atomic scientists and cannot reply authoritatively to this assumption. However, is it likely, applying Occam's Razor, that the Americans, Russians, French, Chinese, British, and Indians, and perhaps the Israelis, have entered into a mutual conspiracy to hide a scientific fact from the world?

There are a number of simple scientific errors in Cathie's writings which cast doubt on his basic thesis, and which we would like to point out. A glaring error involves Cathie's lack of understanding of meteor crater formation, the distribution of which on the earth comprises an important part of his hypothesis. He states (Harmonic 33, p. 39):

> Another thing that bothers me is the fact that all these craters are so symmetrical. . . . Yet in all cases of the craters, we are asked to believe that the meteors descended vertically, in relation to the Earth's surface, and blasted out these perfectly symmetrical holes. Can the scientists explain this . . . ?

Cathie would like to believe that these are not meteor craters but rather explosions from within caused by defects within his grid system. Science is not asking Cathie to believe that all these meteors descended vertically. Much research has been done on impact craters. Here is what a recent elementary textbook has to say (Cazeau *et al.*, *Physical Geology*, p. 425) in the chapter on planetology:

> Laboratory experiments on angle of entry also explain the nearly perfect circular craters we see on the moon. If the angle of entry is greater than 15°, circular craters are always formed.

These results also apply to impact craters on Mars and the earth.

Cathie also says (p. 79) that the number of serious earthquakes has increased by 400%. Cathie does not say where he got this figure. It is a matter of record that over a period of many years there have been serious quakes on an average of one every two to three weeks throughout the world. Only a few of these occur in areas of heavy population and thus make the news. In some years more of the serious quakes may strike the populated areas, but the average number per year has not increased as dramatically as Cathie states. Perhaps more to the point, efficiency in reporting quakes and promulgation of the effects has increased considerably.

Cathie believes volcanoes are caused by electrical and magnetic disturbances. This would be news to vulcanologists who are fairly sure it is due to the rise and extrusion of magma at the surface of the earth. Cathie also believes that diorite is the hardest rock known (p. 142). It is by no means the hardest rock known. Any quartz-rich rock, of which there are many, would be harder because diorite is rich in plagioclase feldspar, which is softer than quartz. Again, Cathie asserts that archaeologists are "completely baffled" by the ruins of Baalbek in Syria (p. 144). Baalbek is known to be of Greco-Roman origin. Archaeologists are not "completely baffled."

Unfortunately, the loose handling of these and other facts calls into question the validity of Cathie's theories.

Despite all of the voluminous writings advancing the ETH, no one has produced, to our knowledge, any convincing artifact from these alien spacecraft or parts of these spacecraft. Nor has anyone presented the world with a genuine alien, dead or alive, despite assertions that several spaceships actually have been shot down through military activity.

## Life Elsewhere and Earth Visitation

Is there life elsewhere in the universe? Especially intelligent life? For many decades these questions have been a source of abiding interest, not only for science-fiction writers, but for scientists too (and many science-fiction writers are also scientists, by the way). The question is germane to our discussion of UFOs because if no life exists elsewhere, then UFOs, flying saucers, or whatever you wish to call them, cannot be alien spaceships. On the other hand, if intelligent life can exist elsewhere, and actually does, then there is always the remote (?) possibility that some UFOs are intelligently directed spacecraft from elsewhere visiting the earth.

Early in the 20th century, the prevailing theory for the origin of the earth and the solar system was that of a close encounter between our sun and a passing star. The near miss produced tidal effects that catapulted solar matter into orbit around the sun, and this matter condensed later into planets. As such, the solar system was treated as a freak of nature resulting from a highly improbable event. It followed that planets were rare, and so life itself was rare. We were alone in the universe.

This idea changed. The new data of astronomy and astrophysics suggested a return to a modified version of Laplace's 18th-century nebular hypothesis. According to this hypothesis, stars are formed as condensations from whirling clouds of interstellar matter. Significantly, planetary formation may be a normal by-product of such stellar evolution. If so, planets are common. And life may therefore be common, assuming life will appear when certain conditions obtain, such as planets of the right size and distance from a star, the presence of water and a rocky crust, and so on. Thus, it has become quite acceptable for a scientist to speak of extraterrestrial life.

If intelligent life is common, even in our own "local" galaxy, still, many scientists point out formidable difficulties in the establishment of actual physical contact.

Among those scientists who have indulged in the "numbers game" of how many civilized planets there are in our galaxy, Carl Sagan has come up with 1 million possible stars with planets harboring advanced civilizations. It would seem that this estimate makes outer space contact a virtual certainty, or at least highly likely.

But wait a minute. When placed in the perspective of the size of our galaxy—80,000 to 100,000 light years in diameter and 10,000 light years thick—the process of one civilization's seeking out another is like looking for the proverbial needle in a haystack. Dr. Sagan calculates that each world would have to launch 10,000 exploratory vehicles per year in order for there to be one spaceship reaching earth each year. Also bear in mind here that our planet would be only one out of a million intelligent worlds of interest to alien peoples.

Some sensationalist writers claim we are now under scrutiny from other worlds because we have suddenly developed atomic power and space travel capability. Sagan answers this assertion as follows (*UFO's—A Scientific Debate*, p. 269):

To imagine that there is something absolutely fantastic, you see, about what is happening right here goes exactly against the idea that there are lots of civilizations around. Because if there are lots of them around, then the development of our sort of civilization must be pretty common. And if we're not pretty common then there aren't going to be many civilizations advanced enough to send visitors.

One aspect of the UFO phenomenon that has always bothered us is the great variety of sizes and shapes and their apparently different methods of propulsion. There is also great variation in the physical descriptions of the UFO occupants, ranging from large, hairy creatures to diminutive humanoids in space suits. This bespeaks more than one alien civilization, perhaps several, that have managed to visit earth. Keeping in mind Sagan's previously mentioned calculations, this seems incredible. This point has been made also by John Keel (in *The Unidentified* by Jerome Clark and Loren Coleman, pp. 183–184) in speaking about the photographic evidence of UFOs:

> . . . although thousands of UFO photos have been taken in the past twenty years, only a dozen or so taken in different parts of the world depict identical objects. If the objects were more uniform in design (and origin), there would now be hundreds of identical pictures. . . . The sightings [are] too numerous and too frequent to be the work of a single technological source.

## Conclusions

In this chapter, we have tended to argue against UFOs being of extraterrestrial origin, although we freely admit that there are many unexplained sightings of a provocative nature. In the absence of natural explanations for these latter cases, we must suspend judgment and accept UFOs, for the time being, as a great mystery of the earth. We do argue against and take issue with fallacious or weak data advanced to support the ETH concept. We agree with Carl Sagan, who concludes (*UFO's—A Scientific Debate,* p. 274):

> I think it's a fact of life that many people are uncomfortable with ambiguity, with a judgment withheld. But, it seems to me, this is precisely where we ought to be on the UFO problem: to say that there aren't enough data, that good judgment isn't possible yet, and that an open mind should be kept. Scientists are particularly bound to keep open minds; this is the life-blood of science.

# PART II
# STRANGE STONE MONUMENTS

# Prologue

What in your opinion is the most prodigious, spectacular architectural phenomenon of the ancient world? The pyramids of Egypt? Stonehenge? Easter Island's monoliths? These are the examples most frequently offered by writers who question human ingenuity as the force behind such extraordinary remains. Some authors, as we have seen, suggest an extraterrestrial origin of civilization, and those for whom such celestrial argosies are too extravagant may fall back on lost continents and global migrations to explain the mysterious prehistoric past.

It is true that, in terms of human labor, considerable physical effort and a triumph of the mind and will are evident in many great archaeological wonders of the world. Enormous weights of stone were involved in their construction. On Easter Island, one of the largest completed images weighs 64 tons. For the outer casing of the Great Pyramid (Khufu, or Cheops) at Giza, limestone blocks were used with an average weight of two and a half tons. At Stonehenge, the inner trilithon uprights weigh between 40 and 50 tons.

Calculations of weight or volume say little about the architectural and early civil engineering skills and creative energy of man, and even less do they illuminate the age-old question of "why?" Let us return for a moment to our earlier question: What is the most extraordinary among all the great constructions of the ancient world? There is an arbitrariness on the part of popular writers who tend to focus repeatedly on the same familiar examples. It is curious that seldom do we think of other works, equally impressive in their monumentality, and yet products of a preindustrial technology.

Man's purposes in life are rarely so heroically portrayed as in religious architecture. The great and lofty Gothic cathedrals of Europe's Middle Ages are awesome and majestic marvels of the human hand, impressing even those of us familiar with modern engineering wonders. The largest religious building

ever constructed is Angkor Wat in Cambodia. Built by Khmer King Suryavar-
man II in the period A.D. 1113–1150 to honor the god Vishnu, the structure
covers 402 acres and its population was at times as high as 80,000.

From what age, then, and from what condition of human life can we
expect to find our most extraordinary achievement in architecture? There is no
single answer. If any answer is to be found it must be looked for in all the
palaces of Persian kings, in cavernous Turkish mosques, in Roman, Byzantine,
Aztec, and Mayan temples. For the archaeologist, what emerges is an inescapable sense of the ingenuity and creative genius of humanity through the ages.
In the astonishing variety of human achievement, the pyramids and henges
appear less and less unique.

## CHAPTER 6

# Stonehenge

*The inhabitants honor Apollo more than any other deity. A
sacred enclosure is dedicated to him in the island, as well as a
magnificent circular temple adorned with many rich offerings.*
— DIODORUS SICULUS

Stonehenge is perhaps the most photographed archaeological site in the world,
yet there is a special quality of mystery about it that no photograph can capture.
No matter how many pictures he may have seen, to the first-time visitor
Stonehenge presents itself as a massive and most impressive spectacle of
beauty and strangeness (Fig. 6-1). One modern writer, Robert Wernick, has
described its crude beauty and power as "strangely disturbing." In his book,
*The Monument Builders,* he writes (p. 145):

> The monument's remote location on an empty plain and its concentric
> rings of megaliths are obviously purposeful. They demand some kind of
> explanation. Yet despite all the fascinating conjecture, Stonehenge will
> probably remain a magnificent mystery—if only because it is the expres-
> sion of an ancient, unfathomable mentality.

Innumerable things have been written about Stonehenge. Some of these
writings, the products of warm imaginations, have contributed little to resolving
the mystery of Stonehenge. Others have produced more plausible theories,
one of which applies astronomical calculations in light of the known history of
the site. We will return later to consider these theories and the purpose of
Stonehenge, but our first objective in this chapter is to address three basic
questions: what is Stonehenge, and when and how was it constructed?

**107**

Fig. 6-1. The site of Stonehenge *(Leonard Smith)*.

Fig. 6-2. Stonehenge seen from the Heel Stone *(Leonard Smith)*.

# A Curious Circle of Stones

Through careful stratigraphic excavations and radiocarbon dating, archaeologists have established a three-stage sequence in the building of Stonehenge.

## Stonehenge I, II, and III

The first construction activity seems to have started about 2800 B.C. Stonehenge I actually began to unfold with the construction of an outer circular ditch and bank that enclosed a ring of 56 pits. The date is based on the radiocarbon analysis of materials found within the bank. On the northeast side, the ditch and bank are interrupted to provide entrance to the inner area. In this earliest phase of construction, a large upright, the Heel Stone, was erected about 100 feet outside the entranceway (Figs. 6-2 and 6-3). The ditch was

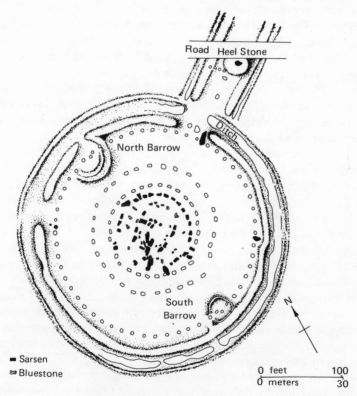

Fig. 6-3. The plan view of the site.

dug into the natural chalky soil, providing the material for the earthwork embankment. The circular bank, measuring 300 feet in diameter, is thought to have been as much as 5 or 6 feet high, although surface erosion has reduced its height to about 2 feet above present ground level.

Another circle, 288 feet in diameter, was formed by the 56 pits, now commonly referred to as Aubrey Holes, named for their discoverer, John Aubrey, a 17th-century investigator of Stonehenge. The holes were roughly circular pits, from 2 to 6 feet across and 2 to 4 feet deep, with flat surfaces at the bottom. Archaeologists found that shortly after they were initially dug out, the holes were refilled with the original chalky rubble and earth. The first refilling of the holes sometimes included cremated human bone. Later some of the Aubrey Holes were dug out again and once again refilled with more cremations, as well as artifacts such as bone pins resembling knitting needles, chipped flint, and so forth.

During this first phase of construction, the presence of the standing Heel Stone must have helped to draw attention to this concentric complex of ditch, bank, and pits. Standing alone in its outlying position and originally enclosed by a small circular earth bank of its own, the Heel Stone is a single large boulder of Sarsen stone (Fig. 6-4). Other surviving stones from later periods at Stonehenge are also identified as Sarsen, a term that refers to a form of natural sandstone that occurs abundantly within about 18 miles of Stonehenge. The Heel Stone stands 16 feet high and weighs approximately 35 tons. It is the only one among all the Sarsen stones that was not deliberately shaped. It is a large monolithic boulder which, as some observers have pointed out, was probably selected for its size and shape, the latter subrectangular in cross section and having a bluntly pointed top.

The Heel Stone was named from a legend, perhaps of the Middle Ages. The legend tells the story of the Devil who, when angered by a pious friar, threw this great rock, striking the friar on the heel. Others have argued that the name has other associations: for example, with the Greek *helios,* "sun," or perhaps with the Saxon word *helan,* "to hide," presumably referring to the momentary "hiding" of the solstice sunrise behind the great boulder.

The appearance of Stonehenge in Stage II changed with the addition of other standing stones, beginning around 2000 B.C. Within the same earthwork circle, the first of a series of some 80 bluestone pillars were erected in two rows forming a crescent or half circle. "Bluestone" refers to a type of volcanic dolerite having a bluish tint, particularly when wet. A fascinating discovery from the petrological studies of a British geologist, Herbert Thomas, was that the bluestones, some weighing as much as 4 tons, were transported from a source in Wales, about 135 miles northwest of Stonehenge (Fig. 6-5).

R. J. C. Atkinson, a chief source of information on the archaeology of Stonehenge, has considered the logical evidence and commonsense observa-

Fig. 6-4. The Heel Stone as viewed from within the site enclosure *(Leonard Smith).*

tions of possible problems in moving the stones. In his book *Stonehenge,* Atkinson describes the most practicable approach, using boats and possibly rafts along the English coastline and tributary rivers. This route would leave only 24 miles of land transport, along which, he proposes, the stones must have been dragged on sledges. Another route, involving the ferrying of bluestones by way of the River Avon, would have required only about two miles overland.

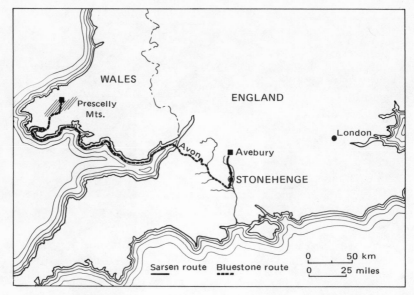

Fig. 6-5. Probable routes of movement of building stones to Stonehenge.

Atkinson reminds his readers that the builders of Stonehenge had no wheeled vehicles or pack animals. Noting that dry-ground sledges are still used on some Welsh and Irish farms, Atkinson did more than theorize about this problem. In *Stonehenge,* he describes an experiment to test the practicability of the method (p. 109):

> A sledge was made to the writer's specification of roughly squared 6 in. timbers, with an overall length of 9 ft. and a width of 4 ft., and the replica of the bluestone was lashed in place upon it. . . . the loaded sledge was then dragged over the down . . . by a party of thirty-two schoolboys, arranged in ranks of four along a single hauling-rope, each rank holding at chest level a wooden bat to whose centre the rope was fastened. It was found that this party could just haul the sledge, weighing some 3,500 lbs. in all, up a slope of about 4° (1 in 15). . . . the use of wooden rollers under the runners of the sledge allowed the hauling-party to be reduced from thirty-two to fourteen, that is by 56 per cent. . . .

Under the best of circumstances archaeology often falls short of revealing historical processes, but imitative experiments such as this help to remove some of the vagueness from theory. Atkinson concluded that at Stonehenge, about 110 men could have moved the Altar Stone, heaviest of the imported stones.

Besides the erection of the bluestones, Stonehenge also grew in Period II by the addition of an avenue that in effect widened and perhaps formalized the entrance to the central part of the site (Fig. 6-3). Almost lost from view today is an extension of the avenue forming a roadway paralleled by two earth embankments, about 50 feet apart. Perhaps meant to be a sort of processional path, the avenue leads from the Avon River to the entrance of Stonehenge. The thought that the bluestones may have been rafted on the Avon, then transported along the track of the avenue, seems a strong possibility in view of the fact that the avenue was laid out along the easiest gradient, carefully selected to avoid slopes or other obstacles in the terrain.

The Heel Stone, although now off the center line of the avenue, not only remains standing but is circled by a narrow and steep-sided ditch. Finally, two additional stones are thought to have stood on the central axis of the avenue.

An even more elaborate level of architectural stonework is displayed in Stonehenge III. Then the site took on its famous silhouette featuring the great Sarsen posts and lintels (Fig. 6-6). This final major construction stage seems to have begun with the removal of the bluestones of Stonehenge II, about 1500 B.C. Next in the sequence was the erection of the five Sarsen trilithons. The trilithons (from the Greek, "three stones") are arches consisting of two 40- to 50-ton uprights supporting a lintel that rests on tenons cut into the top of each upright (Fig. 6-7). The lintels themselves weigh about 10 tons. These gatelike arches form a new horseshoe shape, the open end again pointing toward midsummer sunrise.

Fig. 6-6. The Sarsen Circle *(Leonard Smith)*.

Fig. 6-7. An example of the mortise and tenon construction on lintel and Sarsen upright *(Leonard Smith).*

The new monument of Period III involved bringing more than 80 enormous Sarsen blocks from the Marlborough Downs. When the erection of the five trilithons was finished, the remainder of the Sarsen stone was used to create a large outer circle of the same post and lintel construction. Each lintel was fashioned to a slight curvature and fitted to the next one by a crude tongue-in-groove method. When perched on the columns, they formed a continuous lintel ring encircling the site (Fig. 6-6). In a later substage, some bluestones that had been dismantled for the later rebuilding were reset within the surrounding Sarsen circle.

## The Search for Identity—Who Built It?

There it sits in a pastoral setting, a small but strangely dignified circular complex of earth and stonework features. Someone built it. But who? Over a span of centuries, many ideas have enriched the popular history of Stonehenge, each having its moment of fame. A belief of the Middle Ages tied Stonehenge to the work of magicians. Medieval chronicler Geoffrey of Monmouth tells of Merlin, magician and counselor to King Arthur, and his feat of bringing Stonehenge from Ireland to serve as a burial place. In the middle of the 17th century another writer, John Webb, attributed Stonehenge to the Romans, believing they had erected it as a temple to the sky god Coelus. Eight years later, in 1663, Dr. Walter Charleton, physician to the court of King Charles II, published his view that the monument had been built by the Danes to serve as a place to inaugurate and elect their kings. Charleton wrote, as reported by Wernick *(The Monument Builders,* p. 55):

> Plain stones laid overthwart upon the Tops of the Columns [were for a] convenient and firm footing for such persons of honorable condition who were principally to give their votes at the election of the King.

Although few, if any, 20th-century readers would consider such theories plausible, one suspects that newspapers of the day, had they existed, would have carried a few inches of space for each new assertion and, no doubt, would have made each theory seem entirely credible.

### Druids

There is a mistaken belief that the origin of Stonehenge was connected with Druid worship. The Druids represented a powerful religious sect among Celts of Britain and western Europe. They were known to the Greeks by reputation as early as 200 B.C. and were later described by Julius Caesar during the conquest of Gaul in 58 B.C.

Druid priests seem to have functioned not only as religious leaders but as teachers and judges, and they may also have been a political force among Celtic tribal societies of Gaul and Britain. Although our knowledge of the Druids as a religious body is incomplete, there seems to be no reason to believe that Druids had any part in the origin of Stonehenge, which was already an abandoned ruin centuries before the rise of a Druid Celtic priesthood. Some pertinent factors were reviewed by Atkinson in *Stonehenge* (p. 179):

> The association of the Druids with Stonehenge, now so firmly established in popular imagination, is really of quite recent growth, and its origins lie no further back than the seventeenth century, when John Aubrey first suggested that Stonehenge and other stone circles might be Druidical temples, a suggestion which in the existing state of archaeological knowledge was perfectly reasonable. This notion was taken up and expanded enthusiastically in the eighteenth century by William Stukeley, and popularized in the many guide-books and other works, often based upon his book, which appeared during the following hundred years. Its continuing popularity is doubtless due to its romantic appeal, and not least, I believe, to the fascination exercised on the public mind by the idea of human sacrifice, to which the columns of the popular Sunday press bear eloquent witness.

A dramatic illustration of the Druid link still takes place with the gathering of white-robed Druids at Stonehenge every summer at the time of the June solstice. A ceremonial horn is blown with the arrival of the sun, marking the start of summer in the northern hemisphere. (The modern Druid celebrants have no historical association with ancient practitioners of the cult.)


## Sensationalist Theory

The expectations of some modern writers seem more open-ended and sweeping in their examination of the unexplained Stonehenge. To Richard Mooney, author of *Gods of Air and Darkness,* Stonehenge is a signpost of a long-lost supercivilization. On page 202, he

> . . . supports the view . . . that these sculptures represent one facet of a highly advanced civilization from an unknown epoch.

On page 67, he presents a slightly fuller statement:

> The destruction of a highly advanced civilization in the past by a global disaster would still leave survivors—it obviously has, or we would not be here today. Some of these survivors, the scientists and mathematicians, would wish to estimate the extent of the disaster which had overtaken the planet. To this end they may have devised the mathematical and astronomical complex of which Stonehenge is part.

Other than an implied cause and effect from global disaster to the building of Stonehenge, explanations (if that is the proper term) such as Mooney's are beyond logical analysis since no evidence is presented.

And if we assume, as Erich von Däniken might, that Stonehenge was one of the megaconstructions inspired by the arrival on earth of extraterrestrials, we might still ask, where is the evidence? As we saw in Chapter 4, such revisionist theories would require evidence of abrupt changes brought about by an outside force. But major cultural changes at Stonehenge, as revealed by archaeology, were relatively slow—a long-term growth that spanned more than 1000 years.

## The Work Force

Included in the question of who built Stonehenge is how many men were required and how long it took. We all find ourselves intellectually curious about the answers, but, unfortunately, there are no labor statistics from the prehistoric past. In his book *Stonehenge Decoded* (p. 75), Gerald Hawkins provides a "bookkeeper's" table of minimum work estimates to build Stonehenge I, II, and III. He concludes that a total of 1,497,680 man-days would be required for the physical labor alone, not including administration and logistics.

Using Hawkins's figure, Mooney, in *Colony: Earth* (p. 245), goes further and proposes a population of 75,000 to 100,000 living in the vicinity of Stonehenge for many years. The trouble with any such statistical estimates is that we have no way of knowing how many people were employed. Converting man-days to man-years, we have constructed a table to provide some idea of the time it would take to build Stonehenge for various selected work forces of men (Table 6-1).

**Table 6-1. Alternative Combinations Giving Rise to 4109 Man-Years**

| Number working | Years |
|:---:|:---:|
| 1 | 4109 |
| 2 | 2054 |
| 4 | 1027 |
| 8 | 513 |
| 16 | 256 |
| 32 | 128 |
| 64 | 64 |
| 128 | 32 |
| 256 | 16 |
| 512 | 8 |
| 1024 | 4 |
| 2048 | 2 |
| 4096 | 1 |
| 8192 | ½ |

It can be seen at a glance that this table contains some logical impossibilities. The reader can select from several other possible combinations of men and years, remembering that in this case arithmetic oversimplifies a complex situation. A numerical value of man-days or man-years suggests a kind of unified, continuous plan of construction, but the labor corps at Stonehenge must have varied widely over 1000 years during the stages and substages of rebuilding. The unknown factors of the social machinery at Stonehenge mean that we can still only speculate about the labor requirements.

## Windmills and Beakers

Of course the creation of Stonehenge cannot yet, and perhaps may never, be attributed to any known society. Nevertheless, studies of the archaeology of Britain have given us at least a general answer to the question of who built Stonehenge.

The British Isles, situated on the edge of Europe, received a series of continental influences that affected its cultural development. Sometime prior to Stonehenge I, the idea of agriculture was brought from northwestern Europe by Neolithic, or New Stone Age, people, who also brought with them pottery and animal husbandry. These farming colonists are called by archaeologists the "Windmill Hill" people, a designation that derives from an early earth circle monument—the Windmill Hill site of southern Britain.

The arrival of these immigrants around 3000 B.C. also marks the beginning of the megalithic age for Britain because the new inhabitants made the first use of large boulders, fitted to a variety of uses and forms. One of the inaccurate suggestions about Stonehenge is that it stands alone as a unique and isolated ruin. This is clearly contradicted by the finding of basically similar ruins across Britain, many in the region surrounding Stonehenge itself. Some are monuments of large single upright stones, others are groups of circular or linear alignments, and finally there are roofed chambers or tombs. In The Monument Builders, Wernick summarizes (p. 11):

> Though not every group of farmers and herdsmen in Western Europe
> built monuments, it seems fitting to speak of theirs as the megalith culture.
> And this was a significant time in human history, for it introduced the
> concept of structures consciously designed to last forever.

Although we cannot say what prompted them to do so, it does seem that these late Neolithic Britons were the ones who applied their wisdom and skill to the first construction at Stonehenge.

Even before the close of the New Stone Age, man was aspiring to something better than stone for toolmaking. It came with the discovery that metals, at first gold and copper, could be shaped and used in tools and ornaments. This important technical advantage was brought from Europe by other invaders of Britain, known to archaeologists as the Beaker people. The name derives from their practice of burying the dead with mortuary vessels of pottery in a distinctive "beaker" shape. The contents of their graves suggest that Beaker culture comprised differentiated societies, the upper class depending probably on its close control of trade in metals. The Beaker people introduced gold and copper from Europe, and they may have been attracted to Britain in the first place for the trading, back to Europe, of metals found in Ireland and in western Britain.

Beaker culture also encompassed a kind of megalithic construction. Monuments of upright monolithic stone, sometimes inside earthwork enclosures, were built by Beaker people and are found near their numerous settlements in England. Beaker artifacts found in excavations at Stonehenge lead archaeologists to believe that Period II of that site is the work of the Beaker culture.

## Wessex Culture

Around 1500 B.C., Stonehenge III began. The Wessex culture is the name given to Bronze Age people who continued the tradition of commerce in metals between Ireland and continental Europe. The Bronze Age had dawned with the discovery that tin, when added to molten copper, produces the alloy bronze, much more suitable than copper for tools and weapons. As in Stonehenge II, the people buried their dead in earthen barrows, or mounds, and much of our knowledge of Period III comes from the rich mortuary furnishings of these Wessex tombs. We know, for example, that Wessex society was stratified, with the lower social class or classes led by a kind of aristocracy of warrior-chiefs. Their power and privilege are reflected in the rich variety of tomb contents—ornaments in gold and amber, and weapons and tools of bronze. The catalog includes items from as far away as the Mediterranean.

The picture of Wessex culture gleaned from extensive archaeological study is that of an advanced society, evidently successful in a large sphere of European Bronze Age commerce. With economic wealth and political power at their disposal, the upper-class leaders, perhaps kings or princes, were most likely responsible for the extraordinary Period III monument at Stonehenge. No other known group would have had the required social power to command resources and craftsmanship, and, moreover, Wessex culture burials are found in abundance close to Stonehenge. Departing for a moment from archaeological fact, Atkinson poetically concludes *(Stonehenge,* p. 163):

. . . they [Wessex burials] cluster in great barrow-cemeteries on the crests
of the neighboring downs. . . . When one stands within the stones looking out over their ruins southwards to the barrows on the skyline of
Normanton Down, one can be sure that in them the builders of Stonehenge themselves now rest from their labours.

## Why Was It Built?

We have described some of the bald facts as they have unfolded from
archaeological work at Stonehenge. The problem-solving approach of archaeology has worked well toward answering questions of who, what, and
when. But as we turn to why—those human motives behind man's creation of
a great henge or pyramid—archaeology, by its methods alone, is somehow
inadequate. In the past, prevailing archaeological theory about Stonehenge
viewed it as a sort of sanctuary or temple serving some ritual purpose, perhaps
related to sky worship. That may not be a false view, but since archaeology
cannot recover prehistoric religious motives, it cannot prove its case.

Stonehenge has therefore remained a fertile source of conjectural and
imaginative ideas, some of them very improbable. For instance, there is almost
a compulsion on the part of some people to find alignments between various
kinds of ancient circles, mounds, single buildings, or whole settlements. These
alignments, called leys, are imaginary straight-line paths, which to the ley
hunter represent ancient hidden knowledge of the land. Among the claims for
Stonehenge is the notion that it served as a ley center from which "earth
currents" have emanated. Leys have even been proposed as navigation aids
for UFOs. Scientists find no significance in the theory of ley lines. They argue
that by chance alone many objects or sites of antiquity can be found along a
straight line.

There are other theories that ignore the historical dimensions of
Stonehenge and instead become preoccupied with what lies within its 300-foot
diameter. For example, the "Hidden Halo" hypothesis states that the arrangement of Stonehenge monoliths was influenced by a cloud layer of ice
crystals in the sky. The originator of this theory, I. N. Vail, has proposed in his
book *Canopy Skies of Ancient Man* that a canopy of ice existed for several
thousand years following the last ice age and that early star gazers saw circular
"halos" of ice refraction resulting in patterns that we now see at Stonehenge,
and, according to Vail, in the records of many prehistoric peoples.

## A Digression—Red Halos

Just how prevalent the idea of halos is we cannot say, but the following
remarks from *Stonehenge Viewpoint* were written by Donald Cyr on the

occasion of a visit to a California cave site once occupied by California's Chumash Indians (p. 10):

> When I peered into Painted Cave on this visit, my reaction was one of shock! Suddenly, I could see new meaning in the patterns . . . a HALO in each pattern. I was seeing "Hidden Halos" in an ancient cave. . . . Of course I had seen the same hidden halos in the compartments of New-grange in Ireland. . . . Even the Etruscans had similar drawings on the ceilings of their tombs and in the same shade of red.

In Cyr's view, the special color significance of red seems to apply widely:

> The particular shade of red was meaningful. . . . That proper shade was traditional in this Chumash-land, as it was in Malta, in the land of the Etruscans, and yes, even in ancient Egypt. Had I not seen the same red solar images on the papyrus examples in the British Museum? Of course I had, but seven years ago, when I first visited Painted Cave, I didn't know about "hidden halos" or the red sun of ancient times. In order to get the picture, I had first to visit Newgrange, and Stonehenge, and West Kennet Long Barrow. . . . Only when I could see dozens of "hidden halos" in my mind's eye was I prepared to see the same geometry in California.

Cyr believes that rock drawings of a halo surrounding a solar image means that the Chumash Indians also once witnessed the ice crystals of Vail's Canopy theory:

> The color red means that the shorter waves of light were filtered out so that the sun and the associated halo appeared to be a shade of red. This same color key is used in many lands around the world. The dates when red halos appeared in the sky generally was [sic] prior to about 1500 B.C.

Although some readers might see in the Canopy Theory a promising new line of inquiry, what we have instead is the old familiar pitfall in popular thinking: false analogy, or the inferring of a further degree of resemblance from an observed degree. Donald Cyr's main points seem to be these:

1. An observed degree of circularity.
2. The circularity can be seen at Stonehenge and in pictographs at Painted Cave, California.
3. An assumption that the circles represent prehistoric halo patterns of ice crystals in the sky.
4. Assumption is justified by repeated recognition of halos in prehistoric cultures around the world.
5. Recognition derives from ability to see dozens of hidden halos in the mind's eye.

The circle is a basic shape in nature. It is the shape of the pupil of a human eye, a bubble, or the ripples in a pond. The form of Stonehenge could just as easily reflect the apparent roundness of the horizon as seen from any raised vantage point. And among the free forms of prehistoric pictographs, one of the commonest, if not the most common, is the circle, sometimes spiraled, sometimes as a series of concentric rings. The fact that some of these may represent the sun does not lead to the inductive conclusion that all those artists were observing phases of ice crystal refraction.

And what about the color red? Red pigmentation was in even more common use than Cyr indicates. Natural oxides of iron are found commonly around the world. Such minerals, produced by weathering, are close to the surface and therefore readily available. One such mineral, red ocher, a clayey form of hematite, was widely used throughout prehistory. Cinnabar, a brilliant red sulfide of mercury, was also used. The color red has eye appeal, then as now, for war paint, pottery, pictographs, and so forth. North American Indians of the Archaic period were smeared with red ocher pigment at the time of burial. Prehistoric tribes of Mexico used red earth to stain newborn female babies. South American Indians color their bodies and hair with a red organic pigment.

The use of red mineral paint on some prehistoric pictographs may indeed represent a solar deity. But to conclude, as Donald Cyr does that color and pattern form a chain linking Stonehenge with so-called hidden halos from Egypt to California is a search for the confirmation of a previous conviction. By the misuse of analogy, the finding of simple circles is exalted to a meteorological—red halo—ice crystal theory in which all the elements that confirm a preconception are selected from a mass of experience. This is again the selective "culling" of von Däniken, Mooney, and other sensationalist authors.


## An Astronomical Computer?

In recent years the astronomy of Stonehenge has become the nearly exclusive object of study by some writers. It was noticed, at least as early as the 18th century, that the placement of the Heel Stone and the avenue oriented the site toward midsummer sunrise. Through the 18th and 19th centuries, there were a number of writers who enlarged on the same idea. The use of astronomical knowledge at Stonehenge became an even better established idea in 1965 when Gerald Hawkins, a Boston University astronomer, published his book *Stonehenge Decoded*. Through the use of a computer, Hawkins calculated alignments between features of the site and movements of the sun and moon. He went further with his researches and described the remarkable possibility that the Aubrey Holes served as a computer (pp. 140–141):

> By using them to count the years, the Stonehenge priests could have kept accurate track of the moon, and so have predicted danger periods for the

most spectacular eclipses of the moon and the sun. In fact, the Aubrey circle could have been used to predict many celestial events.

According to Hawkins, by placing markers in the Aubrey Holes and moving them in rotation, one or more hole positions per year (depending on the observations to be made), the priests could keep track of sun and moon cycles and so make their predictions.

Hawkins's deductions about Stonehenge were applied not only to the Period I features, but also to alignments through archways of Stonehenge II, such as when the sun shines through the sunrise trilithon. This dramatic "decoding" led other authors to state that the Stonehenge mystery had been solved. With an attitude leaning heavily toward simplicity, M. H. J. Th. van der Veer and P. Moerman, in their book *Hidden Worlds,* give this concise wrap-up of Stonehenge (p. 125):

> It can be said that one of history's enigmas has been solved with the help of the most modern equipment.

Unfortunately, it is not as simple as that. The astronomical evidence in certain respects is convincing, though in other respects it has been seriously questioned. Elaborating on the work of Hawkins, English astronomer Fred Hoyle agreed that position alignments of Stonehenge I were not random but deliberately and knowledgeably placed to form an ancient astronomical observatory. He disagrees with Hawkins, however, that Stonehenge II and III can also be interpreted as having an astronomical purpose. Hoyle speculates that a major intellectual decline occurred between Stonehenge I and III—a time when perhaps the knowledge of eclipse prediction was lost so that the later setting of stone circles had another undiscovered purpose. In his book, *From Stonehenge to Modern Cosmology,* Hoyle puzzled over it (p. 43):

> The odd and worrying thing is that we have been able to complete the story without referring at all to the later developments at Stonehenge. We might have hoped to find a culmination of intellectual equality in these later constructions, but in vain. Why, if one should succeed triumphantly in predicting eclipses—no easy matter—one should trouble to haul a great mass of stones all the way from the Prescelly Mountains of Wales?

## An Almanac?

If Stonehenge was a sort of record-keeping device, or calendar, we might ask, what need did Neolithic or Bronze Age Britons have for a calendar? Was it used to regulate the timing of agricultural activities? Possibly, but it should be remembered that a precise tallying of days and years was not essential for agriculture. Among peasant farmers, planting and harvesting are seasonal activities, signaled by the natural signs of nature's calendar. The American Indians advised the first European settlers to plant corn according to the size of elm and hickory leaves. In ancient Greece, the cry of migratory cranes marked the time to sow or reap.

As a time-keeping device for cultivation, even Period I of Stonehenge seems larger and more elaborate than necessary. But suppose, as some argue, that Stonehenge served as a religious temple for sky worship in conjunction with its use as a civil calendar device. Knowledge of astronomical phenomena was evidently rather sophisticated if we accept Hawkins's and Hoyle's interpretations of Period I. The knowledge of the appearance of sun, moon, and constellations may have been exclusively in the hands of the priesthood. Such knowledge would have accumulated over many years. The functionaries of Stonehenge may have held their power through the control of this knowledge. Was it those same functionaries who decided that the building material would be stone?

## The Energy of Stonehenge

The use of stone itself has long been one of the site's unanswered questions. We might rationally ask, if Stonehenge was built as an astronomical calculator, why use massive uprights and lintels of stone, shaped to a subtle curvature, when wood was available and easier to work and to transport? One way of accounting for it is to see the stone as embodying a special quality. Janet and Colin Bord in *The Secret Country* describe a practical and powerful purpose for the Stonehenge boulders (p. 19):

> It could be that the scientists of that age, by accurately plotting the positions and relationships in the heavens of the sun, moon, planets and stars, could calculate the optimum time to capture and store the inflow of cosmic energies and the most favorable time for these energies to be released. If this should be so, then the stone circles had four mutually compatible purposes:
>
> 1. Astronomical calculators
> 2. Generators of terrestrial energy
> 3. Storage batteries for both cosmic and terrestrial energies
> 4. Radiating devices to broadcast these energies across the land. . . .

Apart from its use as an astronomical calculator, the Bords' other suggested uses are insupportable. By what evidence can it be established that Stonehenge is a focus of "earth currents"? How would stone circles of cosmic energy be compatible with what is known of Neolithic and Bronze Age peoples?

## The Magic Touch

As we have previously noted, the materials for Stonehenge came from some distance away, in the case of bluestones, evidently all the way from Wales. It has been suggested that these stones, with their conspicuous blue tint, embodied a special mystical or magical property. According to this theory, the Prescelly Mountains are seen as a cult center, perhaps the earthly home of a deity who imparted a special divinity to the bluestone source.

This is only one of the many hypotheses advanced to account for the moving of heavy stones from such a distance. The transport was done by design, not by accident. Clearly, bluestone was a valued material. But could it be that its "mystery" has been overemphasized? The builders might have used bluestone for a material rather than a spiritual quality. Today, architects import Verdi Antique, Vermont, or South African marble for no reason other than for their appearance. Although we are speaking about Stonehenge from a distance of thousands of years, we suspect that often we may be reading into the past some extraordinary qualities that were never there.

## The Path Ahead

In sum, we have looked at several interpretations of Stonehenge. We have seen that facts have often been selectively embellished to explain Stonehenge in a different manner from that which archaeological science would suggest. Yet we have also seen that not all scientific authorities can agree. Obviously, we can approach the meaning of Stonehenge from many starting points. According to the view taken, folk history, archaeology, electro-force fields, or astronomy may prevail. The last is particularly interesting. The astronomy of Stonehenge seems to be a safe middle ground to agree on. The weight of opinion for a long time has favored the idea that Stonehenge did serve, to some extent, as an observatory. Most investigators, both archaeologists and astronomers, would probably agree with the following statement expressed by the authors of a standard archaeological text, Hole and Heizer (An Introduction to Prehistoric Archaeology, p. 411):

> The consensus now is that Hawkins has showed how Stonehenge might have been used; whether it was used that way is another matter. Still the impressive number of correlations between solar and lunar events and the sight lines within Stonehenge is convincing that the monument was planned to do some if not all of the things Hawkins claims.

Notice, however, that astronomical alignments are not considered an overriding and final answer to Stonehenge. In a manner of speaking, whether or not Stonehenge was an observatory is a side issue.

The true mystery of Stonehenge is not the site but its people. The physical evolution of the site is from simple to complex. From its earliest beginnings in Period I, major rebuildings led to the Period III architecture that makes Stonehenge so impressive in the mind and in pictures. But to call Period I people "simple" would be an illusion. They possessed a highly intellectual and technical knowledge and some system of transmitting, over a period of time, knowledge of astronomy, planning, and construction of the monument. Yet they left behind no writing or other so-called civilization arts. This is the mystery that Stonehenge guards so well—the human intelligence that created it.

## CHAPTER 7

# Pyramid

*Du haut de ces pyramides, quarante siècles vous
contemplent.*

[*From the top of these pyramids, forty centuries contemplate
you.*]

—NAPOLÉON BONAPARTE

## Introduction

It stands ten miles west of Cairo, Egypt, on the rocky Giza plateau, a silent
sentinal rising above the waters of the nearby Nile (Fig. 7-1). It is the greatest
single building ever erected by man, a showpiece among ancient mysteries. It is
the Great Pyramid of Khufu (Cheops).

For 4500 years it has stood, already shrouded in antiquity when Caesar's
legions crossed the Rubicon and Christ was born, lived, and was crucified. Its
huge form has been silhouetted against a million and a half Egyptian sunsets.
So huge is it that Napoléon Bonaparte estimated that its 2.5 million stones
could make a wall ten feet high and one foot thick surrounding all of France.
How and for what purpose was this mighty edifice created? What were the
motives of its builders, now long transformed to dust? Why this heroic effort,
such that the Great Pyramid might well stand for another 100,000 years,
perhaps long after human civilization itself has perished?

The Great Pyramid is the subject of this chapter. We want to examine
something of its mystique—the wild and wondrous things that have been said
about it, as well as what archaeologists have learned. But the Great Pyramid is
only one of three that dominate the plateau along the Nile near Cairo. And
these are only three out of a total of about 80 royal pyramids found in Egypt.
And they in turn are just a few of the many other pyramid structures in the
world.

Fig. 7-1. The Nile Valley in flood, looking west-southwest. At the center is the pyramid of Khufu, flanked by Khafra, and in between, one of the queens' pyramids *(Phelps Clawson)*.

Since pyramid architecture was a feature that a number of societies had in common with Egypt, our chapter subject must be widened to include several basic questions. For example, the pyramids seem to have been successful in their purpose, but was their purpose the same everywhere? Also, some writers propose that the pyramids appeared suddenly as part of a ready-made civilization, as claimed by space visitor advocates. Does this fit the conventional explanations of archaeology? And what is the most satisfactory answer to the contention that the Great Pyramid, unlike any other structure on earth, is a unique fountainhead of mysterious and esoteric meaning? Let us begin this tour of pyramids by laying down a few important factual observations.

## Definitions

Pyramids are probably the most instantly recognizable among all ancient artifacts. The architectural description of a true pyramid would be: a geometric structure, usually of stone blocks, brick, or earth fill, with a square or rectangular ground plan, with four sloping sides in the form of triangles that meet in a point at the top (Fig. 7-2). Some pyramids are formed of a series of terraces

Fig. 7-2. Southeast corner of the Great Pyramid. In the foreground is a mastaba of one of the nobles *(Phelps Clawson)*.

with a truncated top instead of sides that meet at the apex. These pyramids have the shape of a platform, often supporting a temple or some elite dwelling.

Ancient engineers recognized in the pyramid the most stable form for grandiose single-building architecture. The center of gravity is only one-quarter of the way up, that is, close to the base in the case of a true pyramid, and one-third of the height of a truncated pyramid. Children with building blocks quickly learn that blocks stacked vertically are very unstable, that only a few can be piled on each other without toppling, whereas blocks stacked like a pyramid create their own stability. Pyramids are known in Egypt, Mexico, India, Asia, Greece, and other parts of the world (Fig. 7-3). But to understand something more about pyramids in the history of various civilizations, we need a further elaboration of architecture.

In the case of pyramids, we are dealing with the single most monumental product of prehistoric technology. Pyramids represent a class of large-scale public works within the sphere of what is generally called civilization. Readers should remember that prehistoric social systems developed in distinctive ways. Many ancient societies never built pyramids or developed major sculptural arts. Their emphasis was perhaps on areas of cultural activity other than those which demand such technology. In these societies, social and spiritual institutions such as old and highly ritualized mythological dances probably served in the

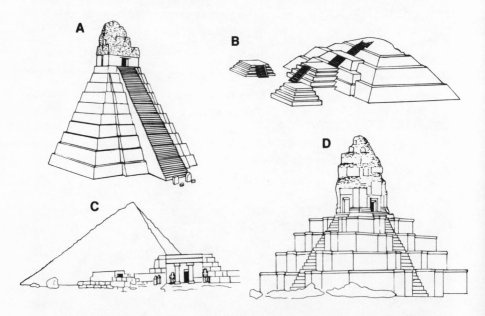

Fig. 7-3. Various types of pyramids: A. Tikal, Guatemala; B. Teotihuacan, Mexico; C. Giza, Egypt; D. Angkor Thom, Cambodia.

main the same societal function as pyramids, that is, to help ensure in some way the continuity of the society. Egyptologists perceive that the pyramid's purpose was to ensure the continuity of the pharaoh in the afterlife, thus most likely contributing to the permanence of Egyptian society.

## Prehistoric Architecture

Some prehistoric societies did make heavy investments of labor in a wide range of architectural types. For instance, the engineering of waterways (man-made canals) is an ancient impulse, and it most certainly involved immense human labor. In the sixth century B.C., the Persian conqueror Darius undertook the construction of a canal linking the Nile with the Red Sea. Collective effort toward massive water control on the Tigris– Euphrates delta was a hallmark of those gifted ancients, the Sumerians. And in the Americas, Aztec civilization developed large-scale hydraulic works. Aztec canals, floodwater irrigation, and floating gardens were objects of great admiration to Spanish conquerors.

Other instances of spectacular architecture are the permanent fortifications that played a central role in warfare prior to the invention of gunpowder. On the European continent and in Britain, Iron Age hill-forts were built with encircling ditches and ramparts. Large forts and castles, often with walls 15 feet thick, enclosed hundreds of acres. The concept of fortifying towns and villages had its counterpart in other places too. Monumental defensive works involving large masonry enclosures or the construction of hilltop fortresses are known in Peru and Mexico. Polynesians living on volcanic islands of the South Pacific terraced commanding hilltops and extinct volcanic craters, fashioning elaborate complexes of terraces, ditches, and palisades.

But unlike waterworks and fortifications, whose practical purposes are obvious, the large imposing pyramids are works of architecture that involve stability and permanence and communicate experience and ideas through form. Archaeologists commonly refer to this as religious architecture—a classification richly represented in history. So abundant indeed are religious relics of the past that it would be difficult to discover major architectural remains that would not be somehow attributable to direct religious significance. To this classification belong the pyramid temples of native American high civilizations such as the Maya, Aztec, and others. The primacy of religion in the Old World was reflected in Mesopotamian temples called ziggurats, in stone circles of prehistoric Britain, and in mosques and pyramids spreading westward from the Pacific borders of Asia to Europe's seaboard. So the pyramids along the Nile are far more than awe-inspiring geometry. They are expressions of a deep seam of complex religious and cultural perspectives.

# The Great Pyramid—Some Popular Views

In this book we have often referred to what we call sensationalist literature. By that we mean the decidedly large corpus of surmises and speculations presented by authors who for the most part have not probed deeply into the specific cases they describe. The romanticized treatment of popular subjects is often superficial, biased, and sprinkled with faulty logic. Often the subject matter is taken out of the framework of conventional knowledge. Writing of this sort does a disservice to the reader in search of objective, neutral inquiries into the seemingly inexplicable.

Egypt, and more particularly the Great Pyramid, can probably lay claim to having inspired more speculation than any other man-made wonder of the past. The speculation has been going on for a long time. The pyramids of Giza are so old that their origins had long been forgotten when the Greek philosopher Herodotus visited Egypt in the fifth century B.C. From our 20th-century vantage point, we can see that Herodotus was sometimes misdirected in his recording of Egyptian history. For example, the notion that the pyramids were built by slave labor is attributed to Herodotus and his Egyptian informants. We now know, based on writings uncovered later, that slavery did not exist in Egypt at that time. Workmen were farmers who were paid in food and clothing for their labor. Yet in other points of detail, Herodotus has been proven correct. As Egyptologist T. G. H. James *(The Archaeology of Ancient Egypt)* says, Herodotus is not a bad source for Egyptian history, only an inadequate one.

The search for the truth of the Great Pyramid later focused on more eccentric explanations, usually centered on biblical interpretations. The 19th-century cult of pyramidology credited Noah with the building of Khufu's pyramid. Although not the first to rely on the Old Testament, Edmond Jeffery *(The Pyramids and the Patriarchs)* gives many justifications for believing that Abraham, forefather of the Hebrews, built the Great Pyramid as an altar to the Living God (p. 144):

> Abram alone had the architects who were capable of planning such an edifice. He alone had the skilled workmen necessary to supervise the thousands of laborers employed upon its construction. He alone had the money necessary to permit such an undertaking, and he alone had the goodwill and co-operation of the Egyptians.

Jeffery makes the further suggestion that Abraham and Khufu were one and the same person. The long line of Egyptian kingship argues against Abraham as a pyramid builder. Jeffery's views are naive and without any foundation of evidence.

The name Charles Piazzi Smyth is prominent among those who have argued that the builders of the Great Pyramid had vast theoretical as well as technical knowledge. Smyth, a Scottish astronomer of the 19th century, developed his own emphatic view that pyramid dimensions, measured in what he called "pyramid inches," demonstrated an ancient knowledge of the wide world and beyond. The pyramid inch, by Smyth's calculation, was equal to 1.001 of our inches. This precise measurement was arrived at most imprecisely, in order to accommodate certain extraordinary claims. Calculate the content of the Great Pyramid in cubic pyramid inches, he said, and you will have the number of all people who have lived on earth since Creation. Multiply its height by $10^9$ and you have the distance from the earth to the sun. Disciples of numerology have stressed the precision of such mathematical measurements which, they say, symbolizes the wisdom of the ages. In his book, *The Past Is Human,* archaeologist Peter White replies (p. 62):

> If one wants to, one can juggle figures in all sorts of ways. For example, the height of the Eiffel Tower in Paris is 29,992 centimetres. This is almost exactly one-millionth of the speed of light (29,977,600,000 centimetres a second). The difference is about 1 in 2000 or five-hundreths per cent. Can this be chance, or were the builders of the Eiffel Tower in possession of this mysterious information? Why would they express this relationship in the tower, built in 1889 for the Paris Exhibition, if it were not to exhibit the marvellously accurate knowledge they possessed.

From among the ranks of occultists, we could add the name of the Englishman W. Scott-Elliot. In a message revealed to him by "astral clairvoyance," theosophist Scott-Elliot proposed mystic recreations of global history, which among other things stated that the pyramids of Giza were built more than 200,000 years ago, in the divine foreknowledge that a great flood catastrophe would cover all Egypt. (The catastrophist view is a theme of endless fascination for some and it is explored in detail in our final two chapters.)

In Richard Mooney's view, expressed in *Colony: Earth,* the Great Pyramid was designed not as a tomb but as a shelter for the safekeeping of knowledge and information. On page 260, he asks:

> Does it not now seem far more likely that the Great Pyramid was designed not as a tomb secure against theft but as a shelter against any combination of forces it may have had to withstand?

A shelter was needed, according to Mooney, for protection against a great catastrophe. Although he does not say whether it was natural or man-made, Mooney describes the catastrophe as the same one that required the building of Stonehenge to measure the effect of the disaster.

From our own professional standpoint, it is difficult to accept such outlandish manipulations of fact and imagination. Archaeological history is made to seem a fragmentary and confused picture with a limited outlook. Of course, free thinking must always be an appropriate part of any scientific inquiry, but we recommend to the reader an active skepticism toward popular writings which ignore facts and solid inference.

## The Pyramid as Miracle Worker

In the marketplace of ideas, one of the newer uses of pyramids goes by the exotic-sounding name of "pyramid power." The premise is simple: power comes from the shape of the pyramid itself, which generates a mysterious energy as yet undefined. *Pyramid Power* (Toth and Nielsen), *The Secret Forces of the Pyramids* (Smith), and *The Secret Power of Pyramids* (Schul and Pettit) are just a few of the books from the full line of titles on pyramid power.

From the reading of such books, we find that the Great Pyramid of Khufu is described as a sort of occult puzzle with amazing beneficial powers. What is promised is that either homemade or commercial scale models will sharpen razor blades, purify water, preserve food, polish jewelry, improve the taste of wine, control pain and heal, make plants grow bigger and faster, and in general aid in everyday practical ways. Also, movie stars and other celebrities have taken to meditating in their own tent pyramids.

We have checked out a few of the claims made about pyramid power's ability to preserve and dehydrate organic material. Eggs, for example, came out of our pyramid after 43 days a smelly, runny yellow, and full of sediment. (However, eggs placed outside the pyramid, interestingly, were in worse condition—black yokes and a strongly obnoxious odor.) Tomatoes in pyramids fared no better than those in brown paper bags. We were unable to sharpen razor blades in our pyramids.

Our own experiments mentioned above were by no means exhaustive. But we are not aware of any systematic scientific investigations that are better. Our results are inconclusive at this juncture. One point does seem indicated, and that is that pyramids do have the power to sell books. The reader is free to conduct his own experiments and make up his own mind about pyramid power.

## Pitfalls

We can suggest several ways in which the public is led to accept superficially persuasive but not necessarily correct conceptions of Egyptian history.

**1. The reader may not realize that the author has either missed or evaded knowledge of the scientific and technical achievements of the ancient Egyptians.** Taking Erich von Däniken's approach to the Great

Pyramid as an example, we find in *Chariots of the Gods?* (p. 78) the following statement:

> Several hundred thousand workers pushed and pulled blocks weighing twelve tons up a ramp with (nonexistent) ropes on (nonexistent) rollers. This host of workers lived on (nonexistent) grain. They slept in (nonexistent) huts which the pharaoh had built outside his summer palace.

Here von Däniken is guilty of factual error and failure to specify. No reasons are given for the statements that ropes and other essentials were nonexistent. In fact, they did exist. To answer this particular point, we turned to a landmark study in Egyptology by Lucas and Harris entitled *Ancient Egyptian Materials and Industries*. In it we find thoroughly documented cases of the finding of rope and cordage that date from both pre-Dynastic and Dynastic periods. Materials were reed, flax, papyrus, grasses, palm fiber, and camel hair.

As for "nonexistent" houses, we have the foundations of barracks situated west of the Pyramid of Chephren. These buildings could house 4000 workmen, according to the discoverer, archaeologist William Flinders Petrie. There are also clay models of houses found in early Dynastic tombs.

**2. Popularizers try hard to convince us that the Great Pyramid is too large to have been built by man.** The outer space propagandists boldly proclaim the presence of alien and superior intelligences who aided primitive man. While admitting that man created the pyramid, they claim that he could have done so only with the help of special devices such as antigravity instruments, laser beams, or radioactive paste for cutting stone. The thought is entrancing, almost irresistible. The melodramatic always is. We are puzzled that no evidence is offered for such visionary notions and that basic archaeological information, as usual, is disregarded. For instance, it is known that from early Dynastic times Egyptians had copper tools—saws and chisels capable of cutting limestone. Discarding the bizarre and unsubstantiated, the reader might invoke Occam's Razor once again and ask himself, why couldn't man have built the Great Pyramid by ingenuity and collective human strength? An objective look at man's works of all ages tells us that we cannot impose arbitrary and quantitative limits on prehistoric man's power and creative knowledge merely because he was prehistoric. He was, after all, *Homo sapiens*.

**3. Radical reinterpretations usually take the Great Pyramid out of its cultural context.** How often have we heard of the "sudden" emergence of Egypt and its pyramids? As von Däniken has put it in *Chariots of the Gods?* (p. 74):

> . . . ancient Egypt appears suddenly and without transition with a fantastic ready-made civilization. Great cities and enormous temples . . . luxurious tombs . . . pyramids of overwhelming size—these and many other wonderful things shot out of the ground, so to speak. Genuine miracles in a

country that is suddenly capable of such achievements without recognizable prehistory.

As we have shown in Chapter 4, sudden appearances are a necessary miracle
in von Däniken's outer space theme. Egypt without a prehistory would indeed
be miraculous. Let us take a moment to outline a few of the salient aspects of
ancient Egypt, particularly the Great Pyramid.

## Pre-Pyramid Age

With some gaps, archaeologists have been able to follow the Egyptians
from the earliest hunting communities of the Stone Age through 31 Dynasties
to the time of Persian domination and beyond. The evidence for beginnings is
usually fragmentary, but we know that around 4000 to 5000 B.C. certain
advances in human knowledge, particularly plant and animal domestication,
were brought to Egypt, probably from that great arc of land in southwest Asia
known as the Fertile Crescent.

This early farming culture is assigned by archaeologists to the pre-Dynastic
period. At that time Egyptians lived in villages, small settled communities of
primitive huts. They hunted and fished to supplement the new agricultural way
of life. They grew wheat and barley and stored it in underground granaries.
They wove baskets and mats and made stone tools and pottery. Detailed
archaeological studies reveal that the latest pre-Dynastic cultures overlapped
Dynasties I and II. For example, types of pottery grave goods with pre-Dynastic
burials were also found in burials of the 1st Dynasty. This cultural continuity
from the pre-Dynastic period contradicts the idea of sudden introductions.

Then prehistory became history with the introduction of writing in the 1st
Dynasty, about 3200 B.C. The Dynasties can be thought of as family groups, the
rulers of which were the kings of Egypt. Since kings were also officially gods,
their exalted positions as monarchs came to be referred to as the "Great
House" (per-o in Egyptian; "pharaoh" in its modern form).

## Tombs

With civilization, man begins to do ennobling (sometimes highly self-
indulgent) things. In pre-Dynastic times, before 3000 B.C., burial practices
consisted of placing the dead in a simple pit dug in the sand. Superstructures, at
first made of mud-brick, were built to protect such graves from the elements
and the possibility of scavengers. These protective superstructures, which have
now come to be known as mastabas, appeared in the 1st Dynasty and they
were the first royal tombs. Through time, the mastaba tomb became more

Fig. 7-4. The excavation of the mastaba tomb of Queen Meresankh III, a grandchild of Khufu, east cemetery at Giza. Ropes are in place for lifting the lid of the sarcophagus *(Phelps Clawson)*.

elaborate (Fig. 7-4). A subterranean burial chamber was placed at the bottom of a shaft that extended down from the superstructure. The superstructure itself could contain one or more rooms, the inner walls inscribed and painted with scenes showing the deceased and his workmen employed in their daily labors (Fig. 7-5).

Fig. 7-5. A relief from an old Kingdom mastaba chapel *(After Breasted,* A History of Egypt).

Egyptologists use the term Old Kingdom to refer to the 3rd through the 6th Dynasties, the age of pyramid building. Saqqara, the cemetery precinct of the ancient Egyptian capital of Memphis, preserves much evidence for the architectural development of the pyramid. The Step Pyramid, for example, built for Djozer, an early king of the 3rd Dynasty, was the first built entirely of stone.

Fig. 7-6. South side of the Step Pyramid at Saqqara *(Phelps Clawson).*

Djozer's great tomb was designed by his highest official, the grand vizier Imhotep, who was himself a man so extraordinarily gifted that he was later deified as a great magician, architect, and physician. The Step Pyramid (Fig. 7-6) was begun as a stone mastaba and was extended to the shape of a tapered pyramid by the addition of six superimposed terraces or tiers.

Another advance toward true pyramid design followed in the 4th Dynasty with construction of the Bent Pyramid at Dashur as a tomb for the sovereign Sneferu, father of Khufu. Also known as the Rhomboidal Pyramid, the Bent Pyramid is unique in having a slope angle that changes about halfway up, the lower half having a steeper inclination. An interesting idea has been advanced as a possible way to explain the curious change of angle. Oxford University physicist Kurt Mendelssohn believes that the "bending" of the Bent Pyramid was linked to an earlier pyramid situated a few miles to the south at Meidum (Fig. 7-7). Egyptologists have attributed to Sneferu both of these pyramids. The builders at Meidum began their work with a step pyramid which was then covered by a massive casing of masonry to give it a truly geometrical pyramidal shape. According to Professor Mendelssohn ("A Scientist Looks at the Pyramids"), a design flaw and subsequent collapse of the entire outer bulk of the Meidum monument during its final building stage persuaded the architect at

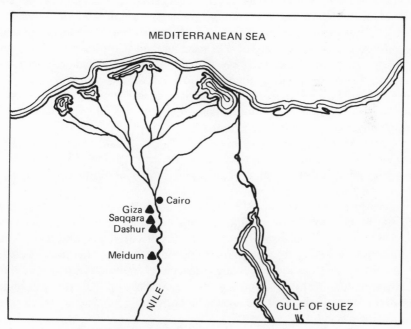

Fig. 7-7. A location map of major sites of the lower Nile.

Dashur to adjust the slope from 52° to a less severe 43° slope, in the hope of avoiding a similar disaster. I. E. S. Edwards, resident pyramid specialist at the British Museum, on the other hand, believes the collapse occurred much later, even perhaps after the 18th Dynasty. Long after its construction, Meidum was described by a visitor as "the very great pyramid." It is unlikely that such a description would be used for a building already in a state of disintegration. Other inscriptions, dating from the later New Kingdom period, found on the walls of a small mortuary temple attached to the Meidum pyramid, lead Edwards to believe that the collapse did not occur for more than 1000 years after its final building phase and may have been caused by an earth tremor ("The Collapse of the Meidum Pyramid"). It is an interesting problem that invites further archaeological testing.


## A 60-Million-Ton Grave

The estimated 2.5 million blocks of stone in the Great Pyramid weigh from 2.5 to 50 tons each. This immense monument whose purpose, as we shall see, was funerary, rises 485 feet above the desert floor. Built primarily of limestone from quarries in the immediate vicinity of Giza, the pyramid was originally covered by casing stones brought from the limestone quarries on the opposite side of the river. Stone for the interior chambers was granite. Undoubtedly the Nile itself was used for the transport, since the source was the principal ancient granite quarry at Aswan, about 600 miles upriver. Construction blocks are fitted together with great accuracy and the sides of the building itself are remarkably accurate in their orientation to north, south, east, and west compass points. The tomb of Khufu (or Cheops as he was known to the Greeks) is the premier example of pyramid architecture, and it symbolizes the unprecedented splendor of the 4th Dynasty.

The plot of Egyptian history thickens, so to speak, following the Old Kingdom. Other periods of outstanding achievement followed but they were interspersed with disintegration of political authority, and from time to time a decline in artistic standards.

We have done no more than sketch pre-Dynastic and Old Kingdom Egypt, but perhaps that is enough to make our point, namely, that what we read in sensationalist digests of Egyptian history about sudden and miraculous appearances is wrong. It seems perfectly clear that the Great Pyramid followed a succession of stages in which early architectural forms evolved gradually to the grand style of the 4th Dynasty. It was by trial and, according to Mendelssohn, painful error that the enterprise of pyramid building achieved such significant success.

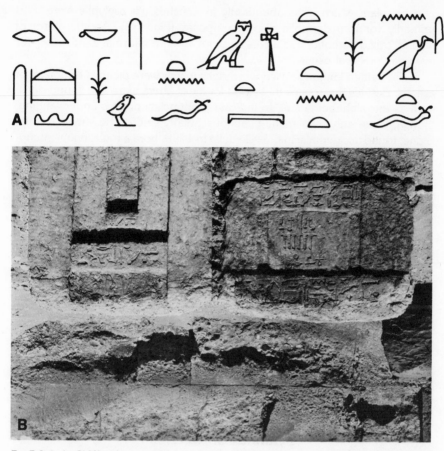

Fig. 7-8. A. An Old Kingdom mastaba text which reads: "It was his mother the King's acquaintance Nedjmet who acted for him while his burial was being prepared." B. The hieroglyphic text as it appears inscribed on the mastaba stone (lower left panel) (*Courtesy Museum of Fine Arts, Boston*).

## The Testament of Archaeology

Early in this chapter, we raised the question of whether all pyramids served the same purpose. This question is closely related to another: How can archaeology be certain of the basic facts of Egyptian history?

A significant point is that the Egyptians recorded their own history. A well-developed writing system was in use by the time of the 1st Dynasty. The writing was carved on wood and stone and written on paper produced from the papyrus plant which grew abundantly in marshes. Hieroglyphs were used primarily for formal public display purposes (Fig. 7-8). Another form of writing, called demotic, was a cursive script, a more informal handwriting for everyday practical use.

The written language used word signs which were pictures of objects or actions. Besides these signs that stood for ideas, there were also sound signs, such as our alphabet letters, to indicate pronunciation. This understanding, though, came only after many years of scholarly study. The famed Rosetta Stone, found accidentally in the Nile delta in 1799, bore a multilingual inscription in Greek, in Egyptian hieroglyphs, and in demotic script. It was chiefly the Rosetta Stone that allowed the brilliant French scholar Jean-François Champollion to decipher the strange writing system through his discovery that the hieroglyphs stood for objects or ideas and for sounds.

The Egyptian concept of immortality is well documented in the data of archaeology. Mummification, elaborate funeral customs, and the burial of clothing, furniture, and food indicate a belief that life after death continued as on earth. Religious texts from the corridors and burial chamber walls of the pyramids leave no doubt about their purpose as mortuary monuments.

Of course, pyramids and royal tombs do not reflect the ordinary existence of the Egyptian masses any more than Mount Rushmore celebrates the achievements of the average American. But various paintings, models, and reliefs provide details of everyday life in Egyptian society: the arts and crafts, transport, trade, social life, food and drink, dress and jewelry, and so forth. All of this gives a picture quite contrary to that of writers who propose mystic origins or sudden derivations from outer space.

## Pyramids of America

Archaeological scholars have followed a lengthy trail of social evolution in America leading from early plant-gatherer remains in dry caves, to the establishment of full-time agriculture at the village farming level, to later periods of Mayan and Mexican civilizations. Specialized pyramid temple mounds were conspicuous parts of the great pre-Columbian centers of classic civilization in both South and Middle America. In Mexico, the Pyramid of the Sun at the great metropolis of Teotihuacan symbolizes the power of an urban civilization that dominated other Mexican and Middle American cultures before its collapse about A.D. 700 (Fig. 7-9). In the forests of Guatemala and Yucatan, pyramid temples tower above Tikal and other principalities of the Mayan civilization (Fig. 7-10). The New World pyramids were constructed by mounding up earth and stone. In the so-called high culture area of Mexico and Middle America, this core of rubble was then faced with a veneer of cut stone masonry. Massive stairways led from a frontal plaza up to the temple. The temple itself was

Fig. 7-9. Pyramid of the Sun, Teotihuacan, Mexico *(Mexican National Tourist Council)*.

constructed either simply of timber and thatch or of masonry. In their final form, both Mayan and Mexican pyramids were covered with thin coats of red-painted stucco. Like their Egyptian counterparts, they were visual spectacles of architectural achievement.

In the short space we have allowed ourselves, let us see how pyramids such as these have figured in the perennial and complex question of American civilization. Traditionally, two main options were presented in the litany of the debate:

**1. Cultural parallels (in architecture and artifacts) between the Old and New Worlds prove the ancient connection.** This is a position still supported by many. As George Carter puts it (Epigraphic Society, February 1975):

> American Indian civilizations were saturated with Old World influences. . . . Transoceanic diffusion occurred, and was important in the growth of civilization in America.

**2. The oceans formed too great a barrier for ancient voyagers, so American civilizations developed in isolation.** This position represents entrenched academic conservatism, but, to many of today's archaeologists, it too is an exaggerated view. There is a growing willingness to concede that voyages, probably accidental, could have occurred. But the question that sometimes leads to overheated controversy is: What effect did such voyages have on the American scene? Could the idea of the pyramid have arrived in America through diffusion over the sea? We cannot be sure.

Fig. 7-10. Temple I, Main Plaza, Tikal, Guatemala *(Stuart Scott)*.

To an anthropologist, a pyramid is a *trait,* meaning a unit or element of culture. Traits can be used to demonstrate connections between people, particularly if several associated traits occur together and consistently. A familiar example is the plow, which does not appear except in association with the draft animal to pull it, some sort of harness, and its human operator. The plow and its component parts spread to many parts of the world as a *trait complex.*

*Fact:*     The pyramid, as a trait, occurs on both sides of the Atlantic Ocean.

*Fact:*     Some large American pyramids also incorporate a tomb as a trait, the tombs containing valuables for the deceased, as in Egypt.

This, to the diffusionist, is a trait complex and presumptive evidence that the pyramid idea was brought to America by ancient transatlantic voyagers. It brings us back to the question, simply stated: How much alike can pyramids be and still be accounted for by parallel or reinvention?

A remarkable resemblance to Egypt can be seen in the well-known case of Palenque, a Mayan site in southern Mexico. There, inside a room of the Temple of the Inscriptions, a stairway was found leading down into the substructure of the pyramid. At a depth of 60 feet, archaeologists found a vaulted chamber and within it the sarcophagus and body of a chief, evidently of extraordinary importance to judge by the richness of the mortuary furnishings of jade and the stucco decorations within the crypt. Does Palenque with its hidden entryway to a tomb constitute proof of external cultural influence? No. One cannot generalize from a single example, as some writers have done with Palenque. Other tombs, much less elaborate than Palenque, have been found within Mayan pyramids, although sometimes they were intrusive, that is to say, they were added at a later time by digging into the pyramid.

By and large, the Maya devoted much more architectural attention to the temples that crowned their pyramids than to burial chambers. The temple consisted of one or more narrow rooms resting on the pyramid platform. Several things point to the use of the pyramid temple as a place of religious ritual. The burning of copal (a resin) incense was considered a holy practice among the Maya. Archaeologists find not only residue from the burning of copal on altars and floors, but occasionally the molded balls of copal prepared for burning. Sealed beneath temple floors are small caches of symbolic offerings: obsidian blades, stingray spines, and other imported objects of ritual significance. The carving of stone monuments and wooden beams in the temples, plus modeled stucco decorations depicting known deities, also point to the use of the pyramid temple as a center of religious authority and the worship of gods of nature: rain, corn, the jaguar, the serpent, and many others.

So the stepped pyramids of America served both as mortuary structures (private tombs) and temples for rites of worship, whereas the Egyptian pyramids did not serve the purpose of civic worship. But still unanswered is the key question: Does the fundamental similarity of tomb-in-pyramid demonstrate ancient contacts from Old World civilizations? The proponents of Old World origins point again to "sudden" appearances. In his book *They Came before Columbus,* Ivan van Sertima postulates a black African–Egyptian presence in America, noting that suddenly, in the "contact" period (800–680 B.C.), stepped pyramids appear in America as a newly introduced culture trait.

From the evidence, or lack thereof, the sudden appearance of American pyramids must remain open to question. But consider the following. Today, major excavations of ancient Mayan ceremonial centers slice through the latest and outermost pyramid temples, uncovering earlier stages in the growth of Mayan architecture. Recently some very early vestiges of Mayan beginnings were discovered from excavations in British Honduras. Archaeologists found both architectural and ceramic evidence of a long period of growth foreshadowing the well-known classic Mayan civilization. Remains of structures with associated human burials more than 4000 years old constitute the earliest Mayan architecture known. Archaeologist Norman Hammond describes it in these words (*Scientific American,* March 1977, p. 125):

> The existence of architectural traditions typical of Classic Maya dwellings,
> such as plastered floors and platforms with timber-framed superstructures
> in the lowlands that long ago is indicative of a developmental period for
> Maya culture of far greater duration than has been supposed.

These discoveries seem to us a broad hint that prototypes of a native American pyramid may one day be found. That would mean that pyramids are not necessarily the result of intercontinental cultural connections. Even if the reader is not prepared to go as far as this, the cardinal question of human inventiveness must still be asked: Does the design and building of a pyramid in one place, at one time in history, preclude the possibility of its reinvention in another place at another time? About 9000 years ago hunters and gatherers in the New World developed agriculture based on their own discovery of plant domestication. The reader should reflect on the possibility that in the thousands of years that followed, the imagination, intelligence, materials, and circumstances needed for the building of pyramids came together more than once in the New World and elsewhere.


## The World of Appearances

In Chapter 1 we saw that the diffusion versus invention issue has had a good working over for many years and that today there is a continuing argument for transoceanic contact on the basis of resemblances in appearance. The argument often relies on traits taken out of their historical context. In *Timeless Earth,* Peter Kolosima gives just such an example (p. 210):

> An extraordinary fact is that the gates of Tiahuanaco (South America) are
> exactly like those of Persepolis in ancient Persia.

There is a likeness, but only in the same trivial sense that prairie houses of

Fig. 7-11. Coral rock trilithon, Tongan Islands, South Pacific.

19th-century America are exactly like those of the 13th-century serfs of Europe because both had walls, roof, and floor and were constructed of sod and stone.

Or take, for example, a great megalithic stone monument hidden in the middle of the South Pacific island of Tongatapu (Fig. 7-11). This prehistoric monument consists of two 30- to 40-ton uprights supporting a lintel that rests on a mortise cut into the top of each upright. Shall we assume that, because of its resemblance to Stonehenge, the Tongan trilithon is the relic of an ancient connection that may illuminate some dark corner of the Stonehenge mystery? Or shall we more reasonably assume that the Tongans, drawing on the inventiveness of the human mind, hit on a remarkably similar but independent variation on a Sarsen trilithon, using simple post and lintel construction?

What we call invention, of course, depends on its definition. The intricate game of chess is not likely to have been invented twice, whereas the use of logs for dugout boats or rafts must have been reinvented many times. The parallel occurrence of pyramids in many parts of the world is a central part of the problem. At the hands of some pyramidologists, pyramids are made to seem mysteriously interconnected, as in this statement from Warren Smith's *The Secret Forces of the Pyramids* (p. 18):

> It is apparent that ancient humans constructed a mysterious network of pyramids throughout the world. We don't know their reasons for building these remarkable stone monuments. We do not know how they built them.

Once again, we would caution the general reader against such superficialities. Archaeology has not revealed any such network and, furthermore, a considerable amount *is* known about how and why the stone monuments were built. Pyramids cannot be treated as isolates, without the archaeological context of each, and, in the same breath, be linked to some vague network of unspecified common history.

## Retrospect

In archaeology, we are really still at the point of only seeing dimly the origins and development of civilizations. It is a question that is broader than pyramids alone, involving as it does art motifs, linguistics, and plant evidence. In the present state of knowledge, no one can offer solid conclusions about the rightness or wrongness of the diffusion argument. As we have seen, pyramids are a major part of that and other controversies. In this chapter, we have tried to recognize that the reader can be easily confused and misled by powerful biases and by what we might call archaeological fiction.

# CHAPTER 8

# Easter Island

*. . . for the whole air vibrates with a vast purpose and energy
which has been and is no more. What was it? Why was it?*

—C. SCORESBY ROUTLEDGE

## An Impending War

Teoso was shaking off sleep that morning. Even the finer things of life on Easter Island, like the thunderous surf, he could not enjoy this daybreak. And his sleep last night had been fitful, all because of the signs of imminent conflict. Days before, his cousins had gone to join a raiding party with others of his people, the Hanau Momoku. Their enemies, the Hanau Eepe, had attacked a fishing party near the statues along the north coast at Petokura. It had been a small attacking force, more like a mob, and the casualties, a few with wounds from stick weapons, were not serious. But the feeling of crisis was serious.

For weeks Teoso had listened to strong and conflicting opinions about the growing hostilities. There had been arguments with the Hanau Momoku over the clearing of land for planting by removal of the surface stone. Long-respected lineage territories had been violated. Teoso knew that in their hearts and minds the Hanau Momoku were greatly troubled. In earlier times, they had shared a peaceful coexistence with those who were now evidently becoming their enemies. Teoso thought to himself that his ancestors, the discoverers of this island, had named it well—Te Pito O Te Henua, "The Navel of the World." But now the island, his world, was threatened by terrible animosities.

It seemed wise to him to look to the defense of his family today, but instead the elders of his kin group insisted that Teoso join the work at the quarry. Slowly and resentfully he made his way to the interior crater of Rano Raraku, the great statue quarry. By now the sun was well above the ocean's horizon on

149

this east end of the island but still low enough in the sky to throw into sharp relief the great effigy images. Here they stood on the lower sloping perimeter of the quarry, and in Teoso's view they appeared powerful, even in their unfinished state, nearly ready to assume their role as deified ancestors.

During the wet and cold winter months, Teoso and his companions had made solid progress in the creation of their statue. On those coldest of days when a small fire was needed for comfort against the southern winds, Teoso would stand warming his hands. From there, his eye could encompass the full size and regularity of features of this idol. Often he pictured the dignity with which this particular image would grace the ahu—a seaside sanctuary already built and awaiting the addition of the statue. Among the carvers of his kin group, Teoso had earned a reputation as a craftsman. He was considered to be very lucky in his job, for as a skilled carver he was exempted from joining the workers. Theirs was the task of lowering the statue from its bed in the stone quarry. Later Teoso would be called on to complete the figure, to give the statue its "sight" by the addition of eyes and to smooth the spinal ridge—a craggy scar left from its final detachment from the quarry.

But all that would come later. He turned again to the immediate task of today's carving—the abrading away of small bits of stone with each blow of the toki, or stone pick. It was a job in which there were no shortcuts, no expediencies. As time wore on that day, Teoso added a worn toki to those that littered the slope below. Almost without stopping, he selected a new one from the basket of quarry tools always available. Even though the image had started to show its anthropomorphic form many weeks ago, Teoso felt a daily disappointment, namely that the progress of one day's sculpturing was scarcely noticeable. It took days of continual hammer-dressing of the block to notice some progress.

And today had been difficult in another way. He had not been able to erase the fear that his family might suffer from the worsening relations with the Hanau Eepe. He remained in that frame of thought all day. It didn't help that today no one from the village came as they usually did to bring fish or fruit for the sculptors' midday meal. It was now midafternoon. Had he been in a position to see, he would have noticed groups of Hanau Momoku men, many armed with obsidian-pointed spears and wooden clubs, filtering east across the lava plain, moving in the direction of Poike. He had seen the large defensive ditch filled with wood and other combustible material which stretched across the Poike peninsula. Unknown to Teoso, large crowds had been gathering near the ditch, the warriors now joined by titled men, elders, even women. Apprehension had changed to terror. Shouting at the top of their voices, some Hanau Eepe warriors seemed to be signaling their combat orders, but fighting tactics appeared to be giving way to pandemonium. Nevertheless, at the sound of a high-pitched blast from a shell trumpet, fires were lit inside and along the full length of the ditch.

At the statue quarry, the day was ending with no more drama than usual except that the carvers were hungry and troubled by feelings of a threatening calamity. The customary eerie quiet fell over this realm of ancestral idols as the quarrymen laid down their tools and prepared to return to the realm of the living. Teoso felt tense and weary as they made their way downslope through the erect but unfinished statues. At that instant, they were joined by a breathless and frantic runner with news of the fighting. For a moment they stood as rigid as the stone images surrounding them. They knew this was more than an intertribal field raid. . . .

The preceding is, of course, an imaginary reconstruction, yet one which rests on archaeological and geological findings, and could be followed to a conclusion on these same grounds. We believe that our fictional character Teoso may have had a real counterpart on Easter Island, living just such an average life. Easter Island has a history, the main outline of which has been determined by archaeology. Popular writers have their own dramatic versions. Here is an example from Erich von Däniken's *Gods from Outer Space* (p. 118):

> A small group of intelligent beings was stranded on Easter Island owing to a technical hitch. The stranded group had a great store of knowledge, very advanced weapons, and a method of working stone unknown to us, of which there are many examples around the world. . . . The unknowns began to teach the natives the elements of speech; they told them about foreign worlds, stars and suns. Perhaps to leave the natives a lasting memory of their stay, but perhaps also as a sign to the friends who were looking for them, the strangers extracted a colossal statue from the volcanic stone. Then they made more stone giants which they set up on stone pedestals along the coast so that they were visible from afar.

Leaving, for a moment, von Däniken's rewrite of history, let us look at Easter Island as it is presently understood, as a background to the contrasting sensationalist literature.

## The Island Rediscovered—A Brief History

The first Europeans to arrive were Dutch. They came in three ships commanded by Admiral Jacob Roggeveen. It was on the evening of Easter Day, April 1722, that the Dutchmen got their first look at the island and named it for the day of its discovery. Observations of the people and conditions on the island were made in only one short day ashore. What the Dutchmen saw and described was limited then by the few hours they allowed themselves on shore, and by an unsettled atmosphere between the natives and the Dutch. When several islanders attempted to rob the Europeans, a struggle developed that unfortunately led to the killing and wounding of several natives.

The Dutch were followed by the Spanish in 1770, and in 1774 famed seaman, navigator, and cartographer James Cook visited the island for several days during his second Pacific voyage. These and other 18th- and 19th-century explorers were the first European eyewitnesses to the mystery of Easter Island. They must have asked themselves the obvious question: Who were these island people? What brought them here and when and from where? And most of all, how could those statues be explained? The Europeans provide us with the first clues to the human history of the island because various landing parties took what notice they could of the people, their dress, their house types, and their apparent customs.

In the list of early European visitors to Easter Island, the successor to Captain Cook was the Frenchman Jean La Perouse, who headed a French government expedition designed to further the work of Cook in Pacific explorations. La Perouse added detailed information on the Easter Islanders whom he observed during a brief ten-hour stay in 1786. In the 19th century, resident French missionaries recorded in their diaries and journals the Easter Island conditions of their time.

When natives boarded his ship at Easter Island in 1774, Captain Cook and his Tahitian translator recognized their speech as similar to that on other South Sea islands they had visited. Estimates of population, based in part on native traditions, indicate that before European discovery the islanders may have numbered 3000 to 4000. The island was annexed by the Republic of Chile in 1888 and Spanish is therefore spoken by many residents. But in the present population of about 1800, those of Polynesian descent also speak the original island dialect of Polynesian.

Although their language has survived, there is little left of the rest of the original culture. Religion is Roman Catholic, based on the gospel instruction of early missionaries. The only village, Hangaroa, consists of European-style houses of wood, concrete block, and corrugated metal. Overall authority is administered by the Chilean government and 20th-century technology has arrived with electricity, a money economy, even television.

Planned archaeological work began in 1914 with a survey of the island carried out by the British Museum. The expedition leader, Mrs. C. Scoresby Routledge, made valuable observations of the images and other major archaeological relics. And in 1934, a combined French and Belgian expedition, led by ethnographer Alfred Métraux and archaeologist Henri Lavachery, researched both the surface archaeology and what the islanders themselves had to say about their historical development.

Emphasizing that many questions still remained unanswered, a Norwegian archaeological expedition under Thor Heyerdahl carried out extensive excavations during five months in 1955–56. In the years that followed, archaeological investigations along with restoration and repair of various structures have continued, in recognition of the outstanding importance of the island as a great outdoor museum. This work has proceeded under the auspices of the Republic of Chile and the joint field direction of William Mulloy of the University of Wyoming and Gonzalo Figueroa of Chile. The application of conventional archaeological science has given us a story of the colonization and human history of Easter Island—one that differs from the loose speculation of sensationalist writers.

## Mystery Island

Time was a resource for nature in creating this volcanic island. Easter Island and similar oceanic islands were raised from the great depths of the ocean floor millions of years ago. Although there is abundant surface evidence of its violent eruptive history, geologists classify its volcanoes as extinct, and they note the essential stability of the island, both during and long before the arrival of its human inhabitants. Three extinct volcanoes are the chief points of reference in the physical nature of the island. The largest and tallest of these, Mount Terevaka, 1700 feet above sea level, together with the other two major craters, Poike and Rano Kao, gives the island its triangular shape (Fig. 8-1).

The landscape contrasts sharply in some respects with other islands of Polynesia. Easter Island is close to the boundary of the tropics and in winter, from May to September, there may be frequent rains and cold winds. The rolling volcanic hills are covered with short grass. The surface lavas, which elsewhere are often effectively covered by lush vegetation, are very much in evidence on Easter Island. In many parts of the island walking is difficult because of the litter of lava blocks. There are few trees.

Accounts of Easter Island have ranged from the rational to the absurd. It seems remarkable that a tiny island, 14 miles long by 7 miles wide and known to Europeans for more than two centuries, should continue to hold its popularity for lovers of the unexplained. Easter Island has become an indispensable chapter in books that attempt to reveal mystical histories of man. Why is this so? Some reasons are easy to guess. For one thing, isolation gives the island a special mystique. Located more than 2000 miles from the South American coast and 1400 miles from the nearest inhabited island, it is one of the world's loneliest places. Its solitary situation in the deserted eastern Pacific Ocean can perhaps have been fully appreciated only by its first discoverer–settlers, who arrived by canoe. Any written account of Easter Island, whether scientific or romantic, has not just an island for its setting, but rather the entire Pacific Ocean.

It has been said that every visitor views the island in highly emotional terms. Some have called it barren and forbidding. Others, like Routledge *(The Mystery of Easter Island)* found in it a wild beauty (p. 255):

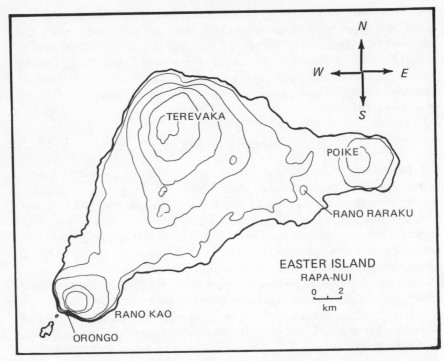

Fig. 8-1. Map of Easter Island.

The whole position is marvelous, surpassing the wildest scene depicted in romance. Immediately at hand are those strange relics of a mysterious past; on one side far beneath is the dark crater lake; on the other, a thousand feet below, swells and breaks the Pacific Ocean, it girdles the islets with a white belt of foam, and extends, in blue unbroken sweep, til it meets the icefields of the Antarctic. The all-pervading stillness of the island culminates here in a silence which may be felt, broken only by the cry of the sea-birds as they circle around their lonely habitations.

Even in the 20th century, the island remains a tiny isolated niche. Its remoteness is a striking feature that has helped to provoke numerous fanciful interpretations of its settlement and history. Of course the primary factor of importance is the presence of the island's renowned statues.

## The Statues

Despite the isolated position of Easter Island, almost everyone knows of its statues, or *moai*, and can recognize their characteristic elongated, brooding faces (Fig. 8-2). They once stood, backs to the sea, on large rectangular platform altars called ahu (Fig. 8-3).

Without the statues, Easter would simply be another Oceanic island of the eastern Pacific. We should focus then on the statuary, asking the same questions that we would in confronting the pyramids, Stonehenge, and any other example of ancient megalithic or architectural art: Who built them? What are they made of? What do they represent? Why are they there? Once again, as usual, we see that there are no completely satisfactory answers. Nevertheless, archaeological field work, together with the careful study of island traditions, has unwrapped much of the aura of mystery surrounding the island and its famous statuary.

Fig. 8-2. Elongated heads and half-buried bodies of colossal stone statues *(Stuart Scott)*.

Fig. 8-3. Author Scott with William Mulloy at Ahu a Kivi, a ceremonial center built about A.D. 1500 *(Robert Koll)*.

## The Stone: Carving and Transporting

Visitors to the island can see, in the quarry, many unfinished examples of statues in various stages of completion. These statues did not magically appear fully completed as is often implied by popular writers. Roggeveen's inspection of the statues led him to the conclusion that they were

> formed out of clay or some kind of rich earth and that small smooth flints had been stuck on afterwards. . . . [*Haklyut*, 1908, p. 15]

We know now though that the megaliths were quarried from a compacted volcanic ash comprising the steep walls of Rano Raraku, a satellite crater near the eastern end of the island (Fig. 8-4). The question of working the stone was given one interpretation when von Däniken wrote in *Chariots of the Gods?* (p. 91) about the ability

> to carve these colossal figures out of the *steel hard volcanic stone* with rudimentary tools. . . . [emphasis added]

By no means steel hard, the Rano Raraku volcanic tuff can be worked without great difficulty using the toki, made from a harder andesitic basalt. These toki, or hand picks used as carving tools, are scattered over the lower slopes of the quarry where they were discarded by the carvers as they became blunted.

Once freed from their rock bed in the quarry (Fig. 8-5), the statues were lowered to the talus slope where they stood awaiting transport to the ahu platforms (Fig. 8-6). Near the top of the crater are snubbing holes, carved as places for rope attachments. One of the most imaginative among the many hypothetical absurdities concerning the island's famous images was proposed by Werner Wolff in *Island of Death*. Wolff's notion was that after they were

Fig. 8-4. The body of a statue and its head, returned from exhibit in Japan. In the distance, the high outer walls of the volcanic stone quarry, Rano Raraku *(Stuart Scott)*.

Fig. 8-5. At the high rim of the volcano are two recumbent statues side by side, their backs still attached to the rock *(Stuart Scott)*.

carved, a volcanic eruption might have blown the statues through the air to their final resting places on the ahu platforms!

Less spectacular but nonetheless remarkable means were in fact used by man for the movement and placement of the island's monuments. The hundreds of gigantic statues around the perimeter of the island were moved from the volcano area, most likely on wooden sleds pulled along prepared roadways still visible today. Heyerdahl, during his investigations, employed demonstration by archaeological experiment, similar to tests described by Atkinson at Stonehenge (see Chapter 6). A crew of modern Easter Islanders fastened a 10-ton statue to a Y-shaped sled of wood and dragged it some distance over the plain without difficulty. As the archaeologists were quick to point out, the experiment does not prove that it was done this way originally, but it does prove that even the largest statues could have been moved by means of direct hauling on sleds.

Heyerdahl's informant was told as a small boy by his grandfather of how the statues were erected on the elevated ahu. In another experiment, a crew of

Fig. 8-6. On the slope below the crater from which they were carved by primitive but conventional means, the great stone images gaze toward their final seaside destinations *(Stuart Scott).*

12 islanders, using long poles for leverage and small volcanic rocks for chocking, gradually raised and erected, by continued levering and building a stone mound of support, a 25-ton statue in 18 days. Once more we emphasize that the experiments do not verify the actual prehistoric techniques, but archaeologists believe that ancestors of the present-day inhabitants did carve, transport, and erect the statues with methods similar to those witnessed by the Heyerdahl expedition. After years of coping with the problems of resetting Easter Island statues, archaeologist–restorationist William Mulloy made the interesting observation (in a personal communication) that given enough men and time, he could move and raise anything, including the largest statue—an enormous and uncompleted figure in the quarry, estimated to weigh 400 tons.

## The Carvers: Could Their Culture Support Them?

Some of the mystery of the island is created by projecting seemingly unanswerable questions. For example, von Däniken makes a statement about food supply in *Gods from Outer Space* to buttress his arguments about assistance from extraterrestrial helpers to create the statues (p. 117):

> So far no one has been able to produce even a tolerably convincing reason why a few hundred Polynesians who found it hard enough to win their scanty nourishment took such pains to carve some six hundred statues.

Really? Little or no food available on this island? ''Scanty nourishment'' is not the way to describe a native economy capable of supporting Easter's population for more than 1000 years. In reality, the waters around Easter Island abound in rich marine food resources including many varieties of finfish, lobster, and shellfish. Add to this the sea birds, such as terns and gannets, that nest on the island, and the assortment of native food plants—taro, yam, banana, sweet potato, sugar cane. The islanders also raised chickens in prehistoric times.

## The Sequence: When Were the Statues Carved?

Enterprising writers, attempting to prove the former presence of a more highly civilized race on Easter Island, have brought settlers as freely as sea birds from China, India, Egypt, and the Lost Continent of Mu, as well as from outer space. But long-term investigations show that Easter Island is Polynesian, settled by those remarkable Oceanists who penetrated the world of the vast

Fig. 8-7. Polynesian ocean-going canoe *(Adapted with the permission of the National Geographic Society)*.

Pacific from their homeland in Asia. They came in sea-going canoes (Fig. 8-7), splendid products of a maritime technology with which they conquered all the major seaways of the Pacific.

It was late in the calendar of human history when the distant corners of the Pacific were settled. A sample of charcoal from the lowest level of one of the ahu gave a date of A.D. 690—the earliest radiocarbon age for Easter Island. But the Polynesians must have made their first landfall on the island before that year. Only after a period of exploring and adapting their life to the new environment would they have begun construction of ahu. These platform altars, built of worked volcanic stone, occasionally were comprised of finely fitted masonry blocks of many tons, but sometimes they were more roughly formed of unshaped stone. The largest were 600 feet long and up to 30 feet high. The ahu served as temples with sloping ramps on the inner side leading toward the worship area and the housing for the priesthood. The ahu temple, as a religious complex, became the specialized form of religious architecture of Easter Island. They also served as burial vaults for the elite. Altogether, about 300 of these sanctuaries can be seen around the perimeter of the island.

The enlargement and remodeling of ahu, and the later practice of carving and erecting statues on them, establish a gradual and important architectural preoccupation. In the words of William Mulloy ("Contemplate the Navel of the World," p. 27):

> Certainly before the time of our earliest radiocarbon date, these islanders had already embarked upon the most remarkable religious building and sculpturing obsession known anywhere in the Pacific.

Both archaeological and traditional (legendary) researches point to two major time periods: an early prehistoric period, followed by a comparatively short era beginning just before European discovery. The statues belong to the early period, which dates approximately A.D. 690 to the 1600s. This was evidently a time of peaceful social order, to judge by the fact that possibly as many as 1000 of the island's famous statues were created then. The legends describe a class-structured society with rule by priests representing religious authority. Sculptors and masons were responsible for various architectural achievements, and food was provided by specialized occupational classes of fishermen and agricultural workers. So the statues belong to the earlier, prehistoric time period. What happened then, according to legends, was a remarkable drama depicted in part at the outset of this chapter—a drama that disrupted the social structure and organization that earlier had sustained the population. In a manner of speaking, the island's history follows the rise and fall, literally, of the statues.

Analysis of the best available archaeological data and correlation of the most consistent island traditions show that the later epoch was marked by intergroup conflict and destruction. The real-life counterpart of our character Teoso would certainly have known about the hostilities remembered and perpetuated in legendary history. According to one version told to the ethnographer Alfred Métraux ("Ethnology of Easter Island") by a native informant, the Hanau Eepe, variously translated as "The Long Ears," or, perhaps, "stocky, heavy-set people," were a kin group inhabiting the Poike area (Fig. 8-1). The elements of war seem to have erupted out of arguments with another kin group, the Hanau Momoku, "The Short Ears," a name which is thought by some to translate more correctly as "slender people." The arguments centered on the refusal of the Hanau Momoku to throw surface stone into the sea in order to clear more land for agriculture. This dispute, probably fueled by other incidents that shaped the pattern of local turmoil, led to a great battle along the ditch at Poike, and according to the legend, nearly all the Hanau Eepe were killed. A pincer movement of the Hanau Momoku, around both ends of the ditch, permitted them to get behind the Hanau Eepe and drive them into their own defensive fire . . . the good odor of the cooked meat of the Hanau Eepe rose into the air, concludes the legend. Archaeologists have confirmed that the long ditch across the eastern headland is indeed man-made and excavations in the ditch provided radiocarbon dating to suggest that the battle occurred about A.D. 1680, a scant 40 years before the arrival of Admiral Roggeveen.

The traumatic turn of events seems to have ushered in an era of aggression and further breakdown of the old religious leadership. Summarizing the available evidence, Mulloy gives us this composite view of the great postwar unrest ("Contemplate the Navel of the World," p. 29):

> Very probably because the devastation of war interrupted food production, the established religious aristocracy lost its essentially magical control and the people degenerated into mutually hostile bands. . . . The hitherto efficient economic equilibrium disintegrated. Crops were burned and farmers prevented from cultivating in safety. Though ritual cannibalism may have been present in earlier times, this now became a more practical activity and people were hunted for food. A frequent theme of the legends of this period relates the suffering of fugitives who hid in caves from human predators. The most horrible of atrocities are described.

As for the statues, they, like other elements of the old society, were affected by the new course of events. The supernatural power thought to be embodied in each of the images would probably be reason enough for the carvings to be maintained for a while. But in the face of continuing intertribal wars, the quarry work must have become increasingly difficult. Over 200 unfinished statues can be seen today in the Rano Raraku quarry. The scene is often described as one of sudden abandonment by writers who seize on the idea to dramatize their own conclusions. But the carving may not have ended abruptly. Because of the nature of the product and the raw material, planning and carving was a long-term project. If the work was not suddenly halted, but rather gradually slowed and eventually stopped, the appearance at the quarry would be the same.

For more than a century after the first Europeans arrived some statues were still standing, but the comments of various European visitors reveal a progressive decline of their importance. In the 18th century, exploring parties of the Dutch and Spanish noted that the great images were still ritually venerated, at least to some degree. But as the years passed, the statues and religious structures suffered erosion, symbolically as well as physically. By the close of the 18th century some statues had fallen or were deliberately toppled, possibly to destroy their supernatural power. A universal token of the social turbulence of the times was the intentional toppling of the statues, tipping them forward to fall face down. Not satisfied with simply overthrowing them, the falling images were sometimes decapitated by the placing of vertical slabs to break their necks (Fig. 8-8). By the advent of the missionary period in 1864, no statues remained standing.

Fig. 8-8. Statues deliberately toppled from their platform stands during the historic period of destruction and social disorder *(Stuart Scott)*.

## The Setting: Are the Statues Unique?

The Navel of the World is not actually centrally situated, as the Polynesian name would suggest, but rather it is marginally located at the easternmost edge of the great Polynesian triangle. Yet there are many reasons, historical, archaeological, and linguistic, for believing that its cultural heritage is truly Polynesian. In language, culture, and appearance, its people were found by the earliest European explorers to be similar to those on the other Polynesian islands, from Hawaii to New Zealand. In finer points of comparison, Easter Island is culturally close to its eastern Polynesian neighbors, the Society, Tuamotu, and Marquesas island groups. Easter Islanders spoke a dialect closely related to Marquesan, and artifact inventories such as house types, fishhooks, adzes, and so forth indicate that Easter Island was settled by eastern Polynesians. Again, let us take the most conspicuous example—the statues. Human sculpture in stone is another cultural trait of the Marquesas and Society Islands, in turn quite similar to carvings found commonly in other parts of Polynesia. The earliest sculptured statuary on Easter Island was not the highly stylized form so often pictured, but was somewhat more naturalistic, resembling general sculptural art of other Pacific archipelagos. So in eastern Polynesia we are dealing with a tradition of sculptured statuary, which on Easter Island merely became a larger and more conventional local style, another example of the fitted masonry of certain Easter Island constructions. This precise stonework was considered a priority item in the theory of a South American center of Polynesian origins popularized by Thor Heyerdahl. Even allowing the possibility of some South American contact, important comparisons must again be made with the rest of Polynesia. The detailed literature on Oceanic archaeology shows that fitted stone masonry was an ancient construction technique elsewhere in eastern Polynesia, in Hawaii, and on Tonga in western Polynesia. Many questions about the Pacific's human history are still unanswerable, but we note again that many writers of the popular mystery school prefer to keep their readers ignorant of this interconnection between Easter Island and its Polynesian ancestry. They substitute various other plots, such as this one from Peter Kolosimo's *Timeless Earth* (p. 44):

> It seems likely . . . that the Incas may have reached Easter Island before the Polynesians and been exterminated or driven out by them. If so, the newcomers may have taken over from the Incas the cult of ancestors, thus explaining the famous statues. . . .

We have said that the images of Easter Island represent a localized and elaborated Polynesian tradition, and that in other respects the island is close, if not geographically, at least culturally, to the heart of Polynesia. Yet the island has a character uniquely its own. There are additional ingredients we might mention. For instance, the island had a form of writing consisting of picto-graphic characters inscribed on small wooden tablets. Known as the Rongo-rongo script, it has been carefully studied, though not as yet fully deciphered. Despite attempts to demonstrate its relationship to an ancient script of India, the Rongo-rongo written language appears to be unique to Easter Island.

Some of the characters on the inscribed tablets are also found as relief carvings near the ceremonial village of Orongo at the southwest corner of the island. Perched between the Rano Kao crater on one side and a 1000-foot drop to the ocean on the other side, it is a grouping of 48 houses of stone slab construction, part of an assemblage of important religious and ceremonial practices (Fig. 8-9). At the south end of the house site at Orongo is a cluster of boulders on the sea cliff overlooking three small islets, Motu-nui, Motu-iti, and Motu-kaokao. The boulders contain relief designs, especially figures of men with bird heads, symbolizing an island-wide ceremony held once each year (Fig. 8-10). The rites were a form of competition between tribes. Designated

Fig. 8-9. Reconstruction of the ceremonial village of Orongo *(Stuart Scott)*.

Fig. 8-10. Ritual carvings of the bird-man cult on lava boulders at Orongo. In the far distance is Motu-nui, largest of the three offshore islands from which eggs of the sooty tern were retrieved by swimmers *(Stuart Scott).*

swimmers competed to be the first to bring back from Motu-nui an egg of the sooty tern, which nests annually on the tiny offshore islands. It seems likely that as much as a religion the ceremony was a means of maintaining power and leadership. Métraux, who in 1934 interviewed the nephew of one of the last bird-men, describes its significance with these words ("Ethnology of Easter Island," p. 331):

> The chief whose servant discovered the first egg received the envied title of bird-man. . . . This title, surrounded with tapus [taboos], brought to its

holder and to members of his lineage certain material advantages and vast moral and religious benefits.

The bird-man cult and its annual ceremony was the principal religious festival of the island and it lasted until after the introduction of Christianity. Today only the tiniest shreds of the old religion linger. An interesting example is found in the island's only church, where to the right of the altar is a wooden figure of the Holy Mother, her head crowned with a carved image of the bird-man cult.

## General Considerations

What we are saying is that many notions popularizing the mysteries of Easter Island have been published by authors who seem to feel little responsibility either to science or to their readers. In *Timeless Earth* (p. 43), Kolosimo declares that Easter Islanders had no logs for transporting statues since

the layer of soil covering the island's rocky surface is too shallow to permit the growth of trees.

It is clear from pollen studies that trees did grow, although they may not have been plentiful, and supplied sufficient wood for the construction of canoes, oars, house frames, fires, and timber materials for the transport of statues. The soil today supports the growth of large eucalyptus groves, miro-tahiti (an orange berry tree), fig trees, and so on.

Later (p. 43) Kolosimo adds that among the island's mysteries are:

. . . the arrangement of the statues, recalling the magic circles of Stonehenge . . . the "bird-men" of Easter Island are certainly akin to the fabulous "fire-bird" met with in India, the Americas and the Mediterranean countries—a creature which seems to have been the symbol of one of the mother civilizations of our earth, the mythical Atlantis.

In other chapters of this book we have seen that such effortless global mixing of cultures adds up to precisely nothing in the light of known human histories. Yet Easter Island still guards some secrets.

Why were the statues carved to conform to one generalized and highly distinctive style? What prompted the development of a written language in a part of the world in which verbal memory was the norm? And why did the bird-man cult, a form of religious life, develop to an extent that has no parallel in Polynesia or anywhere else in the world? William Mulloy, drawing on a long Easter Island research career, gives this interesting summary ("Contemplate the Navel of the World," p. 26):

Nevertheless in sharp contrast to the experience of most isolated peoples and for reasons still largely enigmatic, they developed a surprisingly complex culture, including such unexpected symbols of advancement as a written language . . . a class oriented society with enough coercive power to assemble large crews for spectacular public works projects, an organized priesthood . . . an impressive religious architecture . . . highly stylized stone sculpture productive of about a thousand statues some of which weigh hundreds of tons, and the engineering procedures necessary to transport them.

The prime mystery of Easter Island should not revolve around whether man made the statues and, if so, how. The evidence shows that he did and shows how he did it. Instead, the deeper and far more interesting mystery of the island is this: What sparked the creation of the great images and what motivated the people to continue their carving for centuries on this tiny mote of land that lacked trade and communication with the outside world? The founding populations of Polynesia carried certain common Polynesian traits, both biological and cultural. The same people settled wherever land and resources were adequate. But it was only on Easter Island that Polynesian society was so uniquely modified and directed.

From Stonehenge and Egypt to Easter Island, we have seen that at particular times and places, gifted and energetic people expressed parts of their culture through distinctive and lasting monumental works. Our view is that mystery lies not in the explanation of technology, but in the question of what emotional or intellectual qualities, what faiths, beliefs, or creative imaginations, lead societies to express themselves in such diverse ways.

# PART III

# MARINE MYSTERIES

# Prologue

Few are the regions of the earth so steeped in mystery as the depths of the oceans. Men have traveled over restless waters in fragile vessels since earliest times, but not unmindful of the sea's many moods, the power to give of its bounty, and, equally, to take its share of victims into its dark domain.

Modern men react to the sea much the same as did their ancient counterparts—with fear, respect, humility, and above all, wonder. Even today, the phrase "lost at sea" conveys something akin to removal to an alien and unreachable planet. It is not surprising that the sea should be the locus of many mysteries that trouble the minds of men, prompting myth, legend, and often wild speculation.

We will consider two very well-known marine mysteries in the next two chapters. Chapter 9 is an analysis of the Bermuda Triangle, an area of the Atlantic where it is purported that numerous planes and ships along with their crews have disappeared without a trace. Chapter 10 deals with the famous lost continent of Atlantis. The vision of Atlantis being swallowed by angry waters amid convulsions of the earth is quite enough, one would think, to satiate the mind of the most enthusiastic adherent of the fantastic. But did such a thing really happen? What is the evidence? We shall see.

CHAPTER 9

# The Bermuda Triangle

*The sea never changes, and its works, for all the talk of men, are wrapped in mystery.*

—JOSEPH CONRAD

## The Lore of the Triangle

In 1950 the *Sandra,* a 350-foot freighter, sailed from Miami for Savannah. There 300 tons of insecticide were loaded on board destined for Puerto Cabello, Venezuela. Pulling out from that Georgia port, the *Sandra* chunked confidently towards South America. She furrowed the Atlantic waters into rippling channels as she cleared port. It was the last sight of her ever caught by man. Not a trace of her was ever found.

This is the way Adi-Kent Thomas Jeffrey describes (in her book *The Bermuda Triangle,* p. 78) one strange disappearance in the Bermuda Triangle.

The usual limits of the Bermuda Triangle are defined by lines drawn from Bermuda to Miami, from there to Puerto Rico, and then back to Bermuda (Fig. 9-1). The region encompasses 400,000 square miles. It is also known as the Devil's Triangle, the Hoodoo Sea, the Devil's Sea, and the Limbo of the Lost. Such ominous name tags have been employed to describe an expanse of sea and islands where—apparently—ships and aircraft by the dozen vanish for no earthly reason, along with their crews, leaving no traces. Unmanned ghost ships come and go. Strange lights and unnatural fog are seen. Compasses spin wildly and radios refuse to function properly.

Such events smack of the supernatural. Writers hint knowingly of magnetic fields, other dimensions, time warps, sinister whirlpools, and alien beings imposing their influence in this mysterious area. These rather bizarre interpretations can be entertained only if the disappearances of ships and planes have no

171

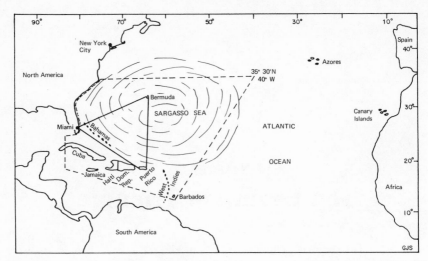

Fig. 9-1. The large area outlined is the sector in which mysterious disappearances occur. The solid triangle is the so-called Devil's or Bermuda Triangle.

logical, physical explanation. The question is, then, what specific peculiar events have occurred here that have given the Triangle its reputation?

## Some Classic Cases

The Bermuda Triangle began to acquire its reputation when there began to appear in magazines and books accounts of the disappearance of the ill-fated Flight 19. Flight 19 was a training mission consisting of five Avenger torpedo bombers (Fig. 9-2) and 14 men. These planes took off in the early afternoon of December 5, 1945, from the Fort Lauderdale Naval Air Station. The mission was to fly a triangular route 160 miles due east out over the Bahamas, then north for 40 miles, and then 120 miles back to base along the hypotenuse of the triangle.

The weather was excellent, the pilots were experienced, and the flight should have arrived back at base in about two hours. But it did not. When the Avengers were at the terminus of the second leg of the flight, and flying north, a strange series of radio transmissions was received from the flight leader, Lieutenant Taylor. He told base he was lost and could not see land. He and his fellow pilots were uncertain which direction was west, and compasses aboard the planes seemed to have gone haywire. Static interfered with communication. Voices of the pilots became panicky. They spoke of the ocean looking "strange" and in one of the final messages they said, "We are entering white water." After that, silence.

Fig. 9-2. These Avenger torpedo bombers are the same planes as those lost on Flight 19 *(U.S. Navy).*

Within minutes, a Martin Mariner search plane (Fig. 9-3) with 13 crewmen aboard was dispatched to look for Flight 19. Unbelievably, it too disappeared. By the next morning, scores of ships and planes were hunting for the missing six planes and 27 men. Nothing was ever found. (We will examine the loss of Flight 19 later in this chapter.)

For many, there was no logical explanation. Even if the planes had ditched in the sea, the men had inflatable liferafts and were trained in ditching procedures. Certainly, some debris should be found. Something extraordinary, something defying natural causes, must have taken place. The last radio messages from the doomed patrol seemed to confirm that something weird was taking place, by the disorientation of the pilots in good weather (couldn't they see the sun in the west?), compasses gone wild, inability to see land, and, perhaps most of all, the cryptic reference to the strangeness of the ocean and "white water." Could the planes and their crews have somehow passed into another dimension? Could they have been kidnapped by UFOs? And what about the Mariner rescue plane? Does it not defy logic that this sixth plane

Fig. 9-3. A Martin Mariner like this was lost while searching for Flight 19. It exploded in midair *(U.S. Navy).*

should also disappear in the same area of sky about the same time? Speculation was widespread, and additional novel theories were offered.

What about other ships and planes? Had others vanished also under similar mysterious circumstances? Some writers wondered and began to dig into the matter, and for them the answer was yes. One author, Vincent Gaddis, referred to the area as the Bermuda Triangle in his book *Invisible Horizons,* and the name stuck. The legend was born.

Researchers into Triangle disappearances were quick to find out about the mysterious case of the U.S.S. *Cyclops,* and also the ghost ship *La Dahoma,* both classic enigmas of the Bermuda Triangle that predated the loss of Flight 19.

The U.S.S. *Cyclops* was a 542-foot ship with a crew of 309 men (Fig. 9-4). How could a ship of such size and with a skilled crew simply vanish? Yet it did, in March 1918, while carrying a load of manganese ore from Brazil to Norfolk, Virginia. We are told the weather was good. Although it was wartime, German records examined after the war show that no submarines were operating in the waters traversed by the *Cyclops* at the time. The big ship carried a wireless. What could have happened so suddenly that a quick SOS could not be sent? No visible wreckage, not even a lifejacket, was ever found.

Can a ship sink beneath the sea in full view of many witnesses and then be

Fig. 9-4. This ore carrier, or collier, is the U.S.S. *Cyclops,* one of the most noted disappearances in the Triangle *(U.S. Navy).*

found days later sailing along without a crew? It is said the *La Dahoma* did. In August 1935 the crew of the sinking *La Dahoma,* a schooner, was rescued by the *Rex,* an Italian ship, and all watched as the schooner slipped beneath the waves. Five days later, the crew of the *Aztec* spotted the *La Dahoma* and boarded it. How could a ship sink and then be resurrected?

In the years following the disappearance of Flight 19, ships and planes continued to be lost. Another ghost ship, the *City Belle,* in 1946; the *Star Tiger,* a British plane, in 1948; and a year later the *Star Ariel,* sister plane of the *Star Tiger.* The list continued to lengthen through the 1960s and 1970s until the cover of one book was claiming that "more than a thousand people . . . over a hundred planes and ships . . . swallowed up into the sea without a trace!" (Richard Winer, *The Devil's Triangle).*

## Major Contentions about the Triangle

We have examined the evidence put forward by those authors who say that disappearances in the Bermuda Triangle are not due to natural causes or human error. These authors emphasize the following major points:

1. There is an abnormally high rate of disappearances in the Bermuda Triangle.
2. All disappearances take place in or very close to the Bermuda Triangle.
3. In virtually all instances the weather was good to excellent at the time of disappearance.
4. Few, if any, of the ships and planes gave any hint of trouble immediately prior to the time they vanished.
5. Despite intensive air and sea searches for the lost vessels, not a trace or a clue was found to explain their loss.
6. There are magnetic forces within the Triangle that cause compasses and radios to malfunction. It is one of two places in the world where compasses point to true north rather than magnetic north.

We think that if there is any validity to any of the above six points, then there is something, indeed, going on in this area that requires more speculative hypotheses than the disbelievers will admit to. On the other hand, should some or all of these contentions be found defective, then the furor over the Bermuda Triangle is probably unwarranted.

We selected for study the 57 cases of lost ships and planes and several derelicts that comprise the hard core of Bermuda Triangle lore (Table 9-1). Let us consider the six points using data from the 57 cases.

## Abnormally High Rate of Disappearances

What is an abnormal number of disappearances per year? The authors we checked mention that there are more than 100 cases, but none actually listed and discussed more than 50 or 60 cases. More than 100 could mean several hundred, but let us assume the number intended is about 200 cases of mysterious happenings in the Bermuda Triangle, and let us restrict the time span to this century (although many authors include losses as far back as 350 years). This works out to be 2.5 ship-aircraft disappearances per year. One author, John Wallace Spencer, does not say that these losses are restricted to the Bermuda Triangle. This could be a reference to worldwide shipping losses. In any event, should we consider 2.5 up to 6.0 losses per year as "abnormal"?

During 1975 the U.S. Coast Guard came to the rescue of 140,000 people. The Coast Guard aided 5600 boats that sank or capsized, 1300 with fires or explosions, and 1000 boats that did not know where they were. Note that this is for only *one year*. It was not an unusual or abnormal year for the Coast Guard. Now if we consider Coast Guard rescue activities for a decade, the number of vessels in trouble from the aforementioned causes is about

### Table 9-1. Losses in the Bermuda Triangle Most Often Cited

| Ships missing | Year | Planes missing | Year |
|---|---|---|---|
| El Dorado | 1502 | Five U.S. bombers | 1944 |
| Genovese | 1730 | Flight 19 | 1945 |
| U.S.S. Grampus | 1843 | DC-3 | 1948 |
| Sandra | 1850 | Star Tiger | 1948 |
| City of Glasgow | 1854 | Star Ariel | 1949 |
| Bella | 1854 | DC-3 missionary plane | 1950 |
| Mary Celeste | 1872 | British transport | 1953 |
| Atalanta | 1880 | Navy Constellation | 1954 |
| Spray | 1909 | Martin P5M | 1956 |
| U.S.S. Nina | 1910 | KB-50 tanker | 1962 |
| U.S.S. Cyclops | 1918 | KC-135 Stratotankers | 1963 |
| Raifuku Maru | 1925 | C-119 cargo plane | 1963 |
| Hewitt | 1921 | Chase-122 cargo plane | 1967 |
| Carol Deering | 1921 | Private plane (Horton) | 1969 |
| Cotopaxi | 1925 | Private plane (Guzman) | 1969 |
| Suduffco | 1926 | Private plane (Cascio) | 1969 |
| John and Mary | 1932 | Private plane (Fischer) | 1971 |
| La Dahoma | 1935 | Phantom jet | 1971 |
| Gloria Colita | 1940 | Private plane (Warren) | 1971 |
| Mahukona | 1941 | | |
| Proteus | 1941 | | |
| Rubicon | 1944 | | |
| City Belle | 1946 | | |
| Driftwood | 1949 | | |
| Southern Districts | 1954 | | |
| Revonoc | 1958 | | |
| Sno' Boy | 1963 | | |
| Marine Sulphur Queen | 1963 | | |
| El Gato | 1965 | | |
| Enchantress | 1965 | | |
| Witchcraft | 1967 | | |
| Scorpion (sub) | 1968 | | |
| Teignmouth Electron | 1969 | | |
| El Caribe | 1971 | | |
| Lucky Edur | 1971 | | |
| Elizabeth | 1971 | | |
| V. A. Fogg | 1972 | | |
| Anita | 1973 | | |

78,000 and the number of people aided almost one and a half million. We hasten to point out that these statistics apply to *all* of the United States coastline and not just the Bermuda Triangle. However, of the 37,000 U.S. Coast Guardsmen, the busiest are those rescuing people in the Bermuda Triangle, one of the great resort areas of the world. Hundreds of thousands of people visit or pass through the Triangle every year. They come by plane and by boat, many of them amateurs without the skill or experience to cope with sudden emergencies. We think that it is a remarkable tribute to the Coast Guard that they let only 2.5 to 6 (?) ships and/or planes slip away from them each year.

## Disappearances Are Concentrated in the Bermuda Triangle

The traditional Triangle is that area enclosed by lines drawn from Bermuda to Miami, from there to Puerto Rico, and then back to Bermuda, as we noted at the beginning of the chapter. Even those authors who stick to this region as the "mystery zone" cite cases that did not take place there, such as the *Mary Celeste*. Of the 57 cases we consider, only 36 (63%) were actually in the Triangle, and 21 (37%) were not in the Triangle. This fact takes away, we think, some of the ethos of the Triangle itself as a sinister area.

Some authors expand the Bermuda Triangle area across the Atlantic to the Azores and as far south as Brazil. John W. Spencer includes the Gulf of Mexico. But most authors agree, it seems, that the true shape of the mystery zone is not really triangular, and each author has his/her own idea of just how big it is. We wonder if one can speak of a zone and a "concentration" of mystery disappearances when the area involved is more than a million square miles.

The larger the area an author claims for the Triangle, the greater the number of ships and planes that can be added to the list. Elizabeth Nichols describes the finding of the *Teignmouth Electron* 700 miles southwest of the Azores. This yacht was skippered by Donald Crowhurst, who had disappeared, and the vessel was deserted. Ms. Nichols, in her book *The Devil's Sea*, noted that *Teignmouth Electron* was located ". . . in the same general area that once shuddered the deserted decks of the *Mary Celeste.*" The *Mary Celeste* did her "shuddering" 1100 miles east of where Crowhurst's boat was found. Neither vessel was in the Bermuda Triangle.

## The Weather Was Good

The oft-repeated statement that in virtually all cases the weather was good is largely untrue. Lawrence Kusche *(The Bermuda Triangle Mystery—Solved)* investigated newspaper accounts of numerous losses in the Bermuda Triangle at the time they occurred, and found that weather was often a major factor, whereas Triangle devotees have a tendency either to avoid mentioning the weather or to say it was good when in fact it was bad.

Here is our summary of 57 cases:

| Good weather prevailed | 15 cases (26%) |
| Bad weather prevailed | 29 cases (52%) |
| Don't know | 13 cases (22%) |

We interpreted bad weather to be hurricanes, gales, and severe storms with high winds and heavy seas.

Let us return to Mr. Spencer, who, in his interesting book *Limbo of the Lost* (p. 5), remarks that

> The major difference between other mysterious sea regions of the world and the "Limbo" zone is that not only are ships engulfed but aircraft as well disappear far beyond the laws of chance.

We wonder how he established the number of losses that would be due to "laws of chance" alone, and what that particular number is? Working out probabilities is fairly simple for things like coins, cards, and dice. The flip of a coin will give you a 50% chance it will be heads because there is only one variable. It is a far more complex matter to derive probabilities of losses in the Bermuda Triangle. Of the many variables to take into account, we include the total number of planes and ships entering and leaving the Triangle, their type and condition, skill of personnel, weather, time of day, and the duration and intensity of the search following a disappearance.

Thus, to speak accurately of the number of disappearances in the Bermuda Triangle as being abnormal or "far beyond the laws of chance" would require input of diverse and complex data.

Why make a mystery out of the loss of the *Anita* in March 1973? The ship was in a severe storm with 45-foot waves. The same storm sank her sister ship, the *Norse Variant,* from which only one crewman survived. The ghost ship *Connemara IV* was found drifting and crewless southeast of Bermuda on September 26, 1955. Another mystery? Maybe, but the yacht was in the path of hurricane Iona with winds up to 180 mph and 40-foot waves. As a final comment, note that the greatest losses are in the stormy winter months (Fig. 9-5).

## No Hint of Trouble

Analysis of our 57 cases produced the following results:

| Complained of trouble | 7 cases (12%) |
| Did not complain | 35 cases (61%) |
| Not applicable | 15 cases (27%) |

The category "not applicable" includes ships lost before the days of radio and a few cases in which we were not able to make a determination. It would seem

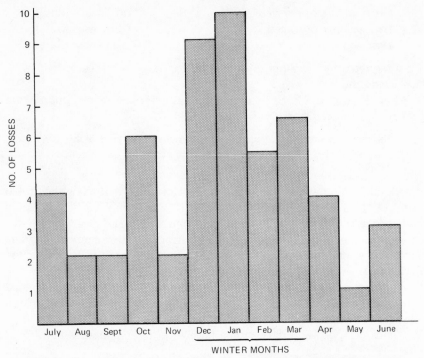

Fig. 9-5. Bar graph showing seasonal relationship to losses in the Bermuda Triangle.

here that we have a point of some validity, with 61% of the disappearances giving no hint of trouble. A possibility is that many of these 35 disappearances might never have happened were the people involved *able* to tell of trouble. This inability may have been brought about by defective or damaged communications (especially in a storm), transmitters with limited range, or poor transmitting conditions. Rescue forces cannot be brought into play unless trouble has been communicated.

However, we are more persuaded by the rapidity with which emergencies arise aboard ships and planes. Recall that in 1975 the Coast Guard aided 1300 vessels with fires or explosions. Violent squalls, arising with little warning, can place a ship in a bad spot in a matter of minutes, especially in the Triangle.

As for planes, let us recount the story told to us by an Air Force major (who wishes to remain anonymous). Some years ago his plane lost a propeller while in flight. The propeller spun off and passed through the fuselage of the aircraft like a buzz-saw, and out the other side, damaging still another engine. Hydraulic lines and wires were cut, disabling most controls. The entire incident occurred in about three seconds. Three seconds—from a well-functioning aircraft to one nearly cut in two, with no controls and two engines lost.

A final point, nowhere mentioned that we could find, is human nature—the pride of a skipper and his crew. Would a skipper relish the idea of having to tell the Coast Guard that he doesn't know where he is? Would not a captain and his crew, in the face of a severe storm or other emergency, be reluctant to admit there was something they could not handle? We think it reasonable that, in some cases, a show of pride may have delayed calling for help until it was too late.

In summary, we think a better, more plausible case can be made for lack of communication (for the reasons given above) than for otherworldly influences.

## Not a Trace Was Found

The expression "without a trace" is a most overworked phrase in the Triangle literature. Sometimes it is otherwise stated: "There aren't even corpses to offer a clue." The thrust of this innuendo is to intimate that traces must be found in all situations where a ship or plane is lost under natural circumstances. Therefore, when traces are not found, the loss must be ascribed to causes unearthly. We are not convinced that this is justified.

In surveying our 57 cases, we find several with traces and clues. We understand traces and clues to mean material evidence such as wreckage, a ghost ship, or distress signals prior to disappearance. It would also be logical to consider a violent storm in the vicinity of the lost craft as at least an indirect clue, since storms do sink ships. Table 9-2 lists 10 losses in which traces were indeed found.

**Table 9-2. "Traces" Found of Missing Craft**

| Ship or plane | Type of "trace" |
|---|---|
| Bella (1854) | Debris, including longboat |
| Raifuku Maru (1925) | Observed sinking in storm |
| KB-50 tanker (1962) | Oil slick |
| Marine Sulphur Queen (1963) | Considerable debris |
| Sno' Boy (1963) | Considerable debris |
| 2 KC-135 Stratotankers (1963) | Wreckage of collision |
| Chase YC-122 (1967) | Both debris and oil slick |
| Scorpion (1968) | Wreck photographed |
| V. A. Fogg (1972) | Entire wreck |
| Anita (1973) | Life preserver |

These 10 losses in which definite traces were later found amount to 18% of all the cases on our list. This hardly squares with the oft-repeated assertion that all or very nearly all of the ships lost in the Triangle vanished without a trace. If we accept a distress call as a clue, and sea-battered ghost ships as a form of material evidence, we can add 17 more vessels for a total of 27 (47%) of the 57. Of the remaining 30 on our list, it cannot be said categorically that there were no traces. More accurately, it could be said no traces *were found.* Wreckage in the open ocean will either sink or be dispersed. The more disturbed the sea surface, as in a storm, the more rapidly this will be accomplished (Fig. 9-6).

Fig. 9-6. Storms at sea such as this do sink ships *(U.S. Navy).*

## Magnetic Forces within the Triangle

The following quote from Adi-Kent Thomas Jeffrey's book *The Bermuda Triangle* is typical of what has been written:

> . . . The magnetic compass doesn't normally point to the true north, but to the "magnetic north." . . . But here we can pause to note a most interesting aspect of this magnetic attraction: there are two places on our globe where the compass *does* point to true north. One spot is the Bermuda Triangle and the other the Devil's Sea area! Is there a relationship here between these compass rarity zones and the disappearances?

The implication here is that within the Bermuda Triangle compasses do not point to magnetic north. The fact is that they do. And also to true north. The compass in this area behaves as it should. The magnetic variation happens to be zero because true north and magnetic north are aligned. If anything, this is an advantage in navigation, as no corrections need be made (Fig. 9-7).

We do not think that the six points discussed above carry much weight as evidence for calling the Bermuda Triangle a mystery area. The ships and planes are in all probability scattered wrecks on the bottom of the ocean in the Bermuda Triangle, and their passengers and crews killed by explosions, accidents, or drowning. Occam's Razor can be applied here.

# Reluctance to Accept Logic

If there is any mystery to the Bermuda Triangle, it seems to be chiefly in the reluctance of some authors to accept perfectly logical explanations for certain losses. For example, Jeffrey reports on the disappearance of the yawl *Revonoc* in 1958. She describes vividly the departure of the yawl with her "sails white and full" sweeping out of the bay at Key West, Florida, and into oblivion. Jeffrey is seemingly unaware of the storm that struck shortly afterward.

Jeffrey mentions the Gulf Stream being "rocked by heavy waves" in a storm with, as Kusche points out, "near hurricane winds from the worst midwinter storm in the history of south Florida." Yet Jeffrey writes, "The 44-foot yawl with all four people aboard had vanished without leaving a trace. It seemed impossible." Impossible? Why? Does she believe that ships cannot possibly sink in a hurricane? Despite her assertion that the yawl disappeared "without a trace" she herself points out that the ship's dinghy washed ashore.

Spencer, in his first book, *Limbo of the Lost*, describes the loss of the *V.A. Fogg*, which sank in the Gulf of Mexico in 1972. The wreck was found and investigated by divers. Spencer pointed out that "the hair-raising part of this mystery is the fact that the captain was discovered sitting in his cabin still

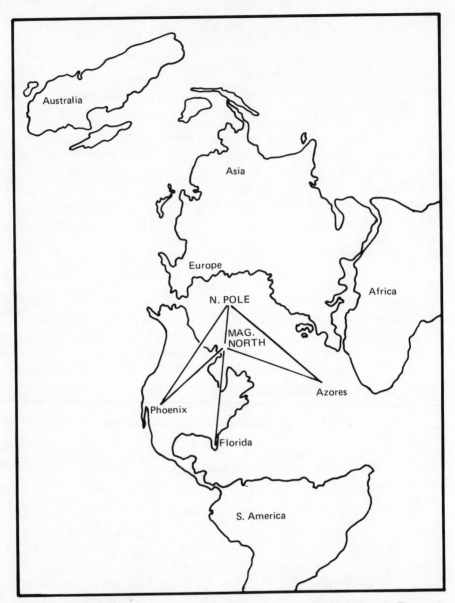

Fig. 9-7. The compass points to both magnetic and true north in the Bermuda Triangle. This is not anything supernatural.

clutching his coffee cup." According to the research of Lawrence Kusche, the divers found the captain's body in the chartroom (not the cabin), but did not report a coffee cup clutched in his hand.

There is no mystery to the sinking of the V. A. Fogg. The wreck was found in 90 feet of water. It had blown up (xylene and benzene were aboard and both are highly flammable). Examination of the wreck showed that this was what had happened. Furthermore, a jet plane pilot flying in the vicinity at the time the V. A. Fogg sank reported a two-mile-high mushroom cloud rising out of the Gulf of Mexico. Yet in the face of these facts, Spencer writes in his subsequent book (Limbo of the Lost—Today, p. 124):

> The truth is that nobody knows what really happened to the V. A. Fogg
> . . . the true story remains in that battered hulk in the depths of the Limbo
> of the Lost.

## Flight 19

What did happen to the flight of the five torpedo Avengers? This is the classic disappearance that established the reputation of the Bermuda Triangle. We will never know exactly what happened, but we do know they were in trouble. They were in radio communication. They said they were lost. There was much discussion. No mystery here. Flight leader Taylor was worried. Being over the open sea, he thought of diminishing fuel and oncoming night.

There is a plausible explanation for what happened. Most Triangle authors do not follow through as persistently as did Richard Winer (The Devil's Triangle), who obtained information from key witnesses such as radioman Melvin Baker, who was in contact with the flight leader during the episode. According to Winer, the flight encountered very strong headwinds out of the northwest that blew the planes south to the Andros Island area. Looking down from his cockpit, the flight leader, Lieutenant Taylor, newly transferred to Florida, confused this area with the southern tip of Florida, for he set an arcing course north and then east so as to approach base from the west. However, this route only brought the planes out over the Atlantic east of the Bahamas, at which point the flight announced it was lost.

According to radioman Baker, who knew where they were, all efforts failed to convince Taylor that his flight was not over the Gulf of Mexico. He kept leading his flight east, hoping to see the west coast of Florida, but in reality he and his doomed planes were flying beyond radio contact, farther and farther east out into the Atlantic. His aircraft, it is further known, would have had to ditch in darkness amid 30-foot waves. The cryptic reference to "white water" could well have been a severe squall disturbing the ocean's surface, as several writers, including Winer, have pointed out. This makes more sense than an alternative theory of entering another dimension.

The Mariner search plane apparently exploded in midair. It disappeared from radar screens at the same time that the crew of the Gaines Mills sighted a ball of flame in the sky. In the past, a number of other Mariners had experienced the same fate as a result of the plane's tendency to leak gasoline fumes.

And what of the *La Dahoma* and the *Cyclops?* While some writers have described how the *La Dahoma* slowly sank beneath the waves, the truth is it was abandoned "in a sinking condition." It continued to remain afloat because of clement weather, even though crewless.

Much has been written about the *Cyclops.* Facts have emerged that militate against the *Cyclops*'s being able to complete her last voyage. The ship carried a heavy cargo, manganese ore, which had been improperly loaded and was subject to shifting. This would enhance the possibilities that the vessel could turn turtle or break in two in heavy seas, tendencies already attested to by former crewmen who had sailed on her. Finally, Lawrence Kusche, searching weather records, found that a severe gale with high winds assailed the Atlantic seaboard on March 10, 1918, at about the time *Cyclops* would have been completing the last leg of her voyage to Norfolk, Virginia. Recent claims that the wreck of the collier was spotted on the ocean floor east of Norfolk remain to be confirmed. If true, this famous sea mystery may be solved.

The *Sandra,* referred to at the beginning of this chapter, also encountered a severe storm with near-hurricane-force winds, as noted by Lawrence Kusche. Our own conclusions concur with those of Kusche, who comments that the Bermuda Triangle is a "manufactured mystery."

## CHAPTER 10

# The Lost Continent of Atlantis

> . . . if Plato had never mentioned Atlantis, someone else
> would have.
>
> —DANIEL COHEN

A legend is usually defined as a nonhistorical or unverifiable story handed down by tradition from earlier times and popularly accepted as historical. It is generally agreed that many legends have some fragments of truth in them. However, the greater the span of time over which such a story is handed down, the greater the opportunity for exaggeration, distortion, and embellishment. This would be especially true in times prior to the invention of the printing press.

The purpose of this chapter is to see if the legend of Atlantis, whose rough outline is known to most people, can survive logical scrutiny. We will also subject the pronouncements and theories of some modern advocates of Atlantis to the same scrutiny.

## The Legend

Prior to 12,000 years ago the Atlantic Ocean was not an open expanse of water. It was occupied by a large island or continent extending from immediately west of the Strait of Gibraltar to the vicinity of the Caribbean Sea (Fig. 10-1). It was a country of varied topography: rugged mountains, broad plains, and river valleys. The soils were fertile and excellent mineral resources were available. An advanced civilization developed on this great continent. The Atlanteans built roads, fortifications, and magnificent temples and palaces. Docks and harbors along Atlantis's shores were used by a sizable navy engaged in vigorous commerce with many nations, especially those bordering the Mediterranean Sea.

187

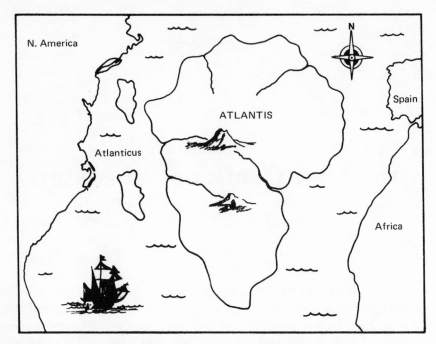

Fig. 10-1. Map of presumed location of Atlantis.

The ambitions of the Atlanteans were expansionist and eventually led to wars of conquest with Mediterranean nations. The Atlantean armed forces were opposed most strongly by the Athenians. The wars were ones of attrition, with the allies of the Athenians falling away one by one. With the wars' outcome apparently still in doubt, and Athens standing alone, the earth was subjected to a convulsion or cataclysm. Earthquakes of great intensity occurred. Lands were thrust downward and giant sea waves swept over the earth. Not only were the Athenians victims of this onslaught, but the Atlantean continent as well. According to the legend, Atlantis sank beneath the Atlantic in a day and a night. Today, its remains can be seen as islands like the Azores and the Canaries, summits of former mountains of Atlantis, protruding above the waters of the Atlantic.

## Plato

The story set forth above was first told by the Greek philosopher Plato, who lived between 429 and 347 B.C. Atlantis is described by Plato in two of his dialogues, *Timaeus* and *Critias*. The latter dialogue was never completed; Plato, in his seventies, apparently turned his attention to other writings and was unable to finish it before death at about age 80. Of course, he may have decided to abandon the project and never intended to finish *Critias*.

Plato never inserted himself in his dialogues. His ideas are voiced through the mouths of others, many of them his contemporaries. A favorite mouthpiece was Socrates. While it may be that Plato wanted to avoid being held accountable for the ideas he expressed by ascribing them to others, he almost always, through his dialogues, projected a zealous love of truth and justice. Yet people of Plato's time grumbled that Plato "put words into their mouths." He did urge kings and rulers to tell a "noble lie" if it were for the good of the people (L. Sprague de Camp, *Lost Continents*, pp. 216–217).

It is probable that Plato would not have lied deliberately to injure someone, or for some other corrupt purpose: This is a surmise drawn from the principles manifest in his writings. Yet he would not hesitate to fictionalize in order to convey a point. He told "parables," for example, in *The Republic*. In this latter case, however, he said he was resorting to fiction. In the case of Atlantis he has Critias state that the topic of Atlantis is genuine history.

In the story of Atlantis as related by Critias, Solon, an ancestor of Plato, journeys to Egypt where he encounters an old Egyptian priest. The priest seems amused that Solon is unaware of the past history of his own country. The priest then recounts the heroics of the Athenians in facing the military might of the Atlanteans. Not only was this flattering to the Greeks, but more importantly, the Athenians of that era felt that they embodied all of the ideal characteristics of Plato's mythical Republic. The historical evidence suggests that Plato intended the story of the war between Athens and Atlantis to be a sequel to his more theoretical account in *The Republic*.

What was Plato up to? Was he a deliberate hoaxer in creating the story of Atlantis? This seems unlikely. If he was not perpetrating a hoax, there are left only a restricted number of possibilities. First, Plato was fictionalizing in conjuring up Atlantis, but through oversight failed to emphasize that his illustration was pure fantasy. This is unsatisfactory in view of the overt statement that Atlantis was "genuine history." Second, Plato was convinced that Atlantis did exist, that indeed there was a historical Atlantis that sank into the Atlantic 9000 years ago. Of course, it may be that Plato was persuaded that Atlantis did exist when in fact it did not. Third, Plato drew upon legends in his time to create his own legend, fragments of historical floods and sinking lands which he utilized to advance his own political and philosophical concepts, using the tale of Atlantis as a vehicle. Let us examine these possibilities.

# Did Atlantis Exist?

## Geologic Evidence

Let us presume that a continent of extensive proportions existed 12,000 years ago in the Atlantic Ocean. Furthermore, let us suppose (as we are told) that this continent supported an advanced civilization with widespread population and building. These people engaged in far-flung trade and war with other nations. Let us accept this premise also, along with the statement that Atlantis sank cataclysmically about 12,000 years ago.

Given these propositions, we should note that they impinge upon both geology and archaeology. Within the realm of geology, we should find some response to the existence of former continents, the cataclysmic sinking of such continents, and the nature of the sea floor in the North Atlantic with respect to its sediments, topography, and objects found on the bottom. In archaeology, we should find some response with respect to the nature, dispersal, and durability of artifacts, or remains of manufactured materials. Also, the data of archaeology should cover the wider intangibles of linguistics, writing, and ideas, philosophy, or lifestyles of the cultures involved.

Geology for more than two centuries has focused its attention on the nature of the earth's outer crust, the origin of the rocks contained therein, and the processes both at the earth's surface and deep within it that have acted to shape and modify the earth. Thousands of observations piled atop thousands of other observations, aided by increasing technological capability to probe the earth and its oceanic depths, have resulted in the amassing of enormous amounts of data. While geologists are quick to point out the areas of the unknown—and the gaps in their knowledge of the earth's history extend back nearly 5 billion years—it must be admitted that the data gathered thus far are formidable for a young science which, at the time of Plato, was rudimentary. (In fact, it was Plato's fellow Greeks who were the first to advance several correct notions about the nature of the earth, and can rightly be numbered among the first geologists.)

The geologic data indicate a thin crust 20–40 miles in thickness consisting of six or seven arcuate plates. According to modern theory, these plates can shift independently of one another, resulting in continental drift. The continental masses ride on these plates and consist of lighter rock material resting on a denser substratum. The present-day continental masses, without exception, are made predominantly of a light rock called granite. The continents of granite can rise and sink gradually. The continental masses are analogous to floating rafts which, being either loaded or unloaded, will ride lower or higher in the water. By gradually, we do not mean rising or sinking in a day or a night (as Plato describes), or even in a span of thousands of years, but more likely over a span of millions of years. Continents have sunk slowly and been flooded numerous times in the 5000 million years of the earth's history, as evidenced by layer upon layer of slowly accumulated marine sediments which now encrust the continents. But the incursions of the sea upon a sinking continent have been so imperceptible that had humans lived during such incursions, they would have gone largely unnoticed within a single human lifetime, or indeed several.

Geology has found no evidence within the geologic past that an entire continent was, with breathtaking swiftness, thrust down into a denser rock substratum. This would blatantly violate the physical laws of nature: It would be analogous to thrusting a custard pie into a denser, unyielding substance such as stiff taffy, and having the taffy yield to the custard pie.

But the supposed time of Atlantis's sinking—12,000 years ago—does not qualify the event in the distant geologic past. Geologic time is measured in millions of years. While numerous geologists indeed study geologic processes going on right now and in the recent geologic past, this is done to facilitate understanding of geologic events of enormous antiquity, long before man appeared on the earth. A geologic event such as the sinking of Atlantis is a geologic "yesterday."

Supposing a continent in the Atlantic had sunk catastrophically 12,000 years ago, what evidence would a geologist look for? Such a profound event would produce enormous effects on the world's shorelines because a sudden continental sinking would create gigantic waves to crash upon even distant shores, and worldwide sea levels would rise virtually overnight. This would be analogous to immersing a large object into a bathtub full of water. The world's shorelines, which have been intensively studied by geologists (Fig. 10-2), reflect no such sudden and catastrophic alteration. Since continental masses are granitic, the center of the Atlantic floor beneath sediment should be granite rock. It is not. It is predominantly basalt, a dark lava rock.

If an advanced Atlantean culture lies beneath the waters of the Atlantic, then there should be remnants of it on the floor of the Atlantic. Even one fragment of an Atlantean house, palace, road, temple, or kitchen utensil would give pause to the most ardent anti-Atlantean. What about the evidence here? Is there any?

The science of oceanography is today very sophisticated and international in scope. Nations such as the United States, the Soviet Union, France, Great Britain, and Japan have sent expeditions into the Atlantic and other oceans to gather data. Scientists aboard these vessels have had no Atlantean ax to grind. They have been in search of other things. Their testimony can thus be considered unbiased. They have been aided in their work by technological skills and equipment developed during World War II and thereafter, such as radar and sonar. In addition, underwater vehicles such as the

Fig. 10-2. Shoreline with raised beach terraces formed when sea level was higher. Such shorelines do not reflect catastrophic changes in sea level *(Buffalo Museum of Science)*.

bathyscaphe, sampling devices, and underwater TV have been developed.

The presumed site of Atlantis has been probed to some extent. Underwater TV cameras lowered to the bottom show no exotic artifacts (Fig. 10-3). Underwater manned vehicles have encountered no Atlantean constructions. They find instead a submarine topography quite unlike the topography of Atlantis described by Plato. Sediment cores retrieved from the bottom show delicate layers of sand, silt, and clay undisturbed during tens of thousands of years of accumulation. A cataclysm would have destroyed and crumpled such layers. These sediments are immensely old, predating the Atlantean episode by far. When asked, one experienced oceanographer, Maurice Ewing, commented that in 13 years of probing and viewing the ocean bottom he had found no trace whatsoever of sunken cities ("New Discoveries on the Mid-Atlantic Ridge"). Unlike eyewitness UFO reports, which are fleeting, some remains of a former civilization should be extant.

The general topography of the North Atlantic is today well known, thanks to continuous profiling by numerous oceanographic vessels of many countries. The most striking feature is the Mid-Atlantic Ridge, a submarine volcanic mountain chain bisecting the North and South Atlantic (Fig. 10-4). On either side it falls away to depths so deep as to suggest that if this ocean floor had been

Fig. 10-3. Sediments of the Atlantic ocean bottom. They show little or no disturbance, nor any Atlantean artifacts.

at one time a subaerial continent, it would have sunk "in a day and a night" to depths of two miles or more. This would require an abrupt and anguished convulsion of the earth not noted by geologists in all of the earth's history. The topography is strangely different. Although the latter is varied, there are few submarine landforms that bear any similarity to valleys, plains, or mountains created by geologic agents on the continents. Were we able to uplift the North Atlantic today above sea level, the continent would in no way resemble the description offered by Plato.

Another major point involves the islands in the North and Central Atlantic which purportedly represent the mountain summits of the submerged Atlantis. Would these not be the very places to which refugees might flee to escape the inundation taking place? They would be the places the Atlanteans might carry their most precious possessions. No one has suggested, as far as we know, that existing structures in the Azores, Canaries, and Madeiras were built by Atlanteans. Nor have artifacts or other personal possessions abandoned by the "refugees" been found on these sites.

In summary, geology offers no evidence that Atlantis ever existed in the North Atlantic, the classical location of this famous continent. There is no evidence in the topography, the sediments, or the rock type. Geologists have

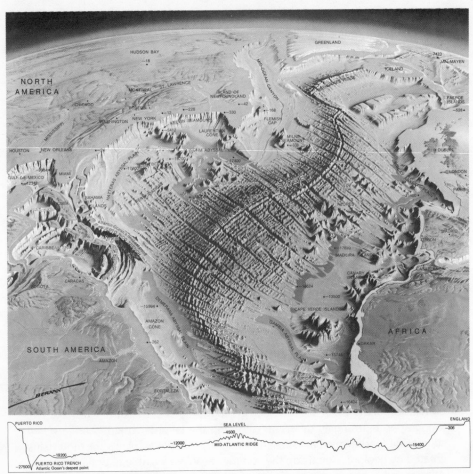

Fig. 10-4. Submarine topography of the Atlantic Ocean floor. Top: The Atlantic Ocean floor, painting by Heinrich Berann (1969). Bottom: Condensed cross-section from Puerto Rico to England, showing comparative depths (*Aluminum Company of America*).

found no traces of former human activity on the ocean floor, nor on the summits of the presumed mountains still lying above the waters of the Atlantic.

On the other hand, it would be dogmatic to say that the geologic search of the North Atlantic has been absolutely thorough. The ocean and its floor are vast. We have directly observed only a small part of it at scattered locations. Something may have escaped the scientists. Furthermore, we think in terms of Atlantean constructions made of solid and enduring stone such as marble or granite. If an Atlantean civilization did exist, conceivably their constructions may have been made of less durable and more perishable materials such as wood or clay bricks. These could have been obliterated in a relatively short time, and thus escaped the inquiring eye of the oceanographer.

We must also admit that we may be looking in the wrong place. Other locations for Atlantis have been suggested (we will return to this question later). In the interim, however, the array of geologic data is negative. The idea is so enticing and appealing, what person, scientist or not, would *not* want to be among the first to discover and announce to the world a scientific treasure of such momentous significance?

## Archaeology

Let us go back to the hypothesis that Atlantis did exist, and let the archaeologist enter. Here is the very person whose life and training are dedicated precisely to the uncovering of such sites as Atlantis. The specific objective of the archaeologist is to discover and uncover (most often by excavation) evidence of past cultures over several millennia. It would be a rare culture indeed that left no evidence whatsoever of its passing. Acting on this premise, the archaeologist searches for, finds, and interprets the various artifacts, inscriptions, monuments, and other remains of past cultures. These are then placed in historical context and related, if possible, to other cultures that coexisted, influenced, or were influenced by earlier or later cultures. The evidence of these interconnections and the level of culture involved might be something as dramatic as the Great Pyramid of Khufu in Egypt, or as humble as a barely recognizable sherd of clay pottery.

While it is true that most archaeological sites are on land, there is a separate subscience of archaeology called marine archaeology. It is, admittedly, a young science, made possible by the development of the aqualung by Jacques Cousteau shortly after World War II. Yet Cousteau and his associates found quick success in their exploration of the floor of the Mediterranean Sea. Ancient Greek and Roman galleys were found. The cargoes of these wrecks, the victims of storms, consisted of wine amphorae, marble statuary, and numerous other artifacts representing the commerce of those ancient times. As yet the marine archaeologist has not applied his techniques to detailed exploration of the Atlantic Ocean floor on the presumed site of Atlantis.

If Atlantis did exist, and did engage in widespread commerce with other nations, especially within the Mediterranean area, then we might reasonably expect that goods were interchanged between Atlantis and other countries. Something tangible, something with the equivalent of "made in Atlantis" on it—pottery, a marble statue, inscriptions in a distinct language, rings, or other decorative objects—would have been dispersed to other civilizations that safely weathered the crisis that obliterated all traces of Atlantis itself. Yet the archaeologist has searched futilely for Atlantis in the remains of Athens and other Grecian sites. There is nothing in the remains of the Babylonians, Sumerians, Egyptians, or others to indicate trade of these peoples with an Atlantis, nor in any of the cultures that foreshadowed these civilizations. Each shows a progressive, systematic, and distinct development without any unknown or "alien" cultural influence that might be ascribed to an Atlantis. Further, not one document unearthed by the archaeologist in cultures of Plato's time or before specifically mentions Atlantis or any reasonable facsimile thereof. Finally, archaeologists can add nothing as a result of excavations on the supposed summits of Atlantis itself—the Azores, the Madeiras, and the Canaries.

In summary, despite the obvious acclaim discovery of Atlantis would bring, neither the archaeologist nor the geologist is able to marshal support for Plato's story.

## Growth of the Legend

In the centuries following Plato's story of Atlantis, writers and armchair speculators often were attracted to the Atlantis theme. Hundreds of books and thousands of articles appeared, expanding upon what Plato wrote. New "evidence" was brought forth concerning the location and character of Atlantis and the origin and degree of development of its inhabitants. Many of these writings are sheer drivel, rejected even by staunch adherents to the historicity of Atlantis. Societies and other organizations sprang up devoted to the Atlantean theme. Many of these societies published journals. High levels of absurdity were reached as extreme groups printed Atlantean money and postage stamps, and, in their meetings, grunted at each other in what they conceived to be the Atlantean language.

To these individuals, proof of Atlantis was quite obvious. There are pyramids in Egypt and pyramids in Yucatan. The survivors, after Atlantis sank, fled east to Egypt and west to Yucatan, taking their Atlantean knowledge with them. This knowledge involved pyramid construction. That the two kinds of pyramids are not that close in details of their construction (see Chapter 7), that they were built in different time periods and for different purposes, and finally, that Mayan writing and Egyptian hieroglyphics are in no way similar—such elementary archaeological facts as these were ignored.

## Location of Atlantis

Earlier in this chapter we pointed out the complete lack of evidence for any continental remains at the bottom of the North Atlantic. At the same time we admitted that, if there were an Atlantis, we might not be looking in the right place. A search of the literature on Atlantis for a logical and viable alternative results in confusion. Berlitz, in his survey of the literature, reports more than 40 suggested locations scattered over the earth. These include such unlikely and unsubstantiated sites as Belgium, Siberia, and East Prussia. The improbability of such sites suggests that Atlantists, lacking scientific guidelines in their search, are groping for any straw that will buttress their cause.

Two sites have gained special favor among Atlantist enthusiasts: (1) Bimini, in the Bahamas, and (2) the island of Santorini (or Thera) in the Aegean Sea.

No one suggests that the island of Bimini was once the nerve center of an Atlantean empire. Rather, it is touted as a western outpost of Atlantis, based on discovery during the 1960s of man-made walls lying today beneath 20 to 25 feet of water. The ruins have not been studied in detail but have been estimated to be about 12,000 years old. The existence of underwater ruins is not surprising to scientists because worldwide sea level was about 300 feet lower when the walls and other constructions at Bimini were built. Such evidence of submerged human construction can be found in shallow nearshore waters all over the world. Sea level has been rising at varying rates during the past 12,000 years. Most recently, the rate of rise has slowed to 4.5 inches per century. Prior to that time, the rate was one foot or more per century. There is thus nothing in the mere presence of these Bimini ruins that dictates conclusive proof of an advanced transatlantic civilization sunk cataclysmically 12,000 years ago.

The island of Santorini (Thera) in the Aegean Sea has been investigated by scientists. There is little doubt from its shape and rock material that the place was subjected to an unusually violent volcanic blast that destroyed what was once a more extensive land area. The date of this eruption is about 1400 B.C. The eruption was responsible for the premature demise of the Minoan civilization on the island of Crete (see Chapter 13).

The volcanic destruction of Thera has been compared by geologists to the eruption of Krakatoa in 1883. Krakatoa, situated in the Java Strait, erupted so violently that at one point during the eruption a cubic mile or more of rock material from the shallow ocean floor was blown 17 miles into the air, and the disaggregated particles were swept around the entire earth. The noise of the explosion was heard 3000 miles away.

In all probability the explosion at Thera was much more violent, affected the entire Mediterranean area, and caused darkening of the sky, earth tremors, and large sea waves to flood coastal habitations. Was Thera, then, the "real" Atlantis? While this cataclysmic explosion may have given rise to legends eventually linked to Atlantis, there are several points that militate against it as the site of Atlantis. Artifacts and extinct communities have been found on Thera, but nothing suggesting a highly developed culture. If one were to insist that Thera was the real Atlantis, then Plato's story must be changed in two respects. First, Plato states that Atlantis was located *beyond* the Pillars of Hercules. Thera lies *within* the Pillars of Hercules. Second, the size of Atlantis must be drastically altered. Plato stipulates that Atlantis measured 2000 by 3000 stadia. A stadium equals 607 feet. This means a continent of about 78,000 square miles. The entire area of the Aegean Sea, in which Thera is but a speck, is only 33,000 square miles. L. Sprague de Camp, noting the changes Atlantists sometimes make in the writings of Plato, states (*Lost Continents,* p. 84):

> . . . you cannot change all the details of Plato's story and still claim to have Plato's story. That is like saying the legendary King Arthur is "really" Queen Cleopatra. All you have to do is to change Cleopatra's sex, nationality, period, temperament, moral character, and other details and the resemblance becomes obvious.

Another location for Atlantis has been suggested by a pastor from Germany named Jurgen Spanuth (see Lytle Robinson, *Edgar Cayce's Story of the Origin and Destiny of Man,* p. 21). He claims to have found, in Egypt, the writings of the old Egyptian priest who told Solon about Atlantis. With these documents, Spanuth says, he found the ruins of a sunken Atlantean fort exactly where the old documents said it had been. This spot is six miles off the coast of Heligoland in the North Sea. Apparently, nobody else has seen these priceless documents of Spanuth.

Heligoland is a small island of 380 acres which is populated by about 1800 people. The island consists of soft red sandstone rimmed by steep cliffs. In past wars, it was a militarily strategic site and was often referred to as the "Gibraltar of the North Sea." A Danish possession early in the 19th century, it was seized by the British in 1890, and subsequently became a German possession and large naval base. It was bombed by the British several times during the Second World War.

It is a historical fact that Heligoland was once a much larger island, but has suffered rapid reduction in size from wave erosion and rise in sea level. In A.D. 800 the island had a circumference of 120 miles. In A.D. 1300 it was 15 miles in diameter and would have included the site of Spanuth's fort. It would appear thus that the so-called Atlantean fort was built much more recently and by someone else. It seems that if Atlantis was not located in the North Atlantic, there is little to support alternative locations.

## Contemporary Writers

During the 1960s and 1970s an upsurge of interest took place in astrology, ancient megaliths, and various occult subjects. A major catalyst for this upsurge was the writings of von Däniken, whose basic premise as we have seen

(Chapters 3 and 4) was that ancient astronauts had visited the earth and altered the evolutionary direction of humankind. Interest increased in Atlantis also, with a number of speculations advanced linking together ancient spacemen and the lost continent.

Let us consider some of the proofs of Atlantis's former existence as supplied by modern enthusiasts. One of these is Serge Hutin. In *Alien Races and Fantastic Civilizations,* he gives due credit to Plato for revealing the existence of Atlantis and quotes extensively from Plato's writings. In this connection, Plato had written that after the sinking of Atlantis ". . . the sea in those parts is impassable and impenetrable because there is a shoal of mud in the way; and this was caused by the subsidence of the island." Hutin then comments on this by stating (p. 56) that "this last sentence would apply very well to the present Sargasso Sea." And then, in a footnote, Hutin states further (p. 148):

> The Sargasso Sea is still imperfectly known, since propeller-driven ships do not venture into it, because of the accumulation of seaweed it contains.

Were Hutin's book written in 1770 rather than 1970, his remarks would be understandable. However, thousands of persons have passed routinely through the heart of the Sargasso Sea in the central North Atlantic (including the authors) without impediment of any kind. True, there is scattered seaweed, but the thick, impenetrable masses of seaweed of many stories are a fiction. Yet even if there were impenetrable masses of seaweed out there, this would not in itself suggest a sunken continent below the Atlantic.

Richard Mooney *(Colony: Earth)* reasons that Plato's description of Atlantis is inconsistent with an empire in the Atlantic and more closely resembles ". . . a Mediterranean type of culture not noticeably different from that readily understood by Solon and Plato" (p. 158). Mooney would like to make ancient Crete the actual site of Atlantis but he recognizes the difficulty of reconciling Plato's dimensions given for Atlantis with the much smaller area of Crete. To get around this, Mooney asks (p. 159), "But what if Solon had erroneously translated the symbol for 100 as 1000? The two symbols in Cretan script are almost indentical [sic]." If this were so, all statistics of size, numbers, and dates would be reduced by a factor of 10 and would seem to make Mooney's thesis credible. For example, the destruction of Atlantis would have occurred 900 years before Plato rather than 9000 years, and thus would coincide with the explosion of Thera and the demise of the Minoan civilization on Crete. Although this arithmetic mistake theory is interesting, Mooney does not offer additional direct evidence that such a mistake occurred. Nonetheless, eight pages later (p. 167) he concludes that "Plato's Atlantis has been shown to be true, even in its details of dimensions." The leap from conjecture to acceptance as fact is not justified. Perhaps to emphasize his point, Mooney asserts (p. 161) that "In 1958, an earthquake in Alaska raised seismic waves more than 1200 feet high, a phenomenon called by the Japanese name tsunami." The error here by Mooney is by a factor of 10. The maximum height ever noted for any tsunami is 120 feet.

Lytle Robinson, a disciple of psychic Edgar Cayce, uses Cayce's readings to prove the existence of Atlantis, with updates by Robinson. For example, on page 22 of *Origin and Destiny* we find that

> Other scientists have recently begun to give weight to the Atlantean theory. Maurice Ewing, of Lamont Observatory and a leading authority on the Atlantic ocean bottom, is one. His expedition found that an abrupt change took place about 11,000 years ago in the Caribbean Sea from cold water type plants to warm water types, and 17,000 years before that there were widespread earth changes. Other U. S. oceanographers, Walter Sproll and Robert S. Dietz of the Environmental Science Services Administration, have theorized that Australia and Antarctica are parts of a once super-continent. They constitute the lands of the continental drift theory—broken off chunks of the present contintents, they believe.

This statement seems to imply that Ewing himself was an Atlantis advocate. He was not, as mentioned earlier in this chapter. The change from cold-water to warm-water plants was occasioned by the warming up of the planet with the end of Pleistocene glaciation. Sproll and Dietz are by no means the originators of the continental drift theory, and in any event the drift they are talking about occurred 200 million years ago and has nothing to do with Atlantis.

A further statement by Robinson that submarine lavas in the Atlantic were found with a glassy structure and therefore "could only have solidified in the open air" is incorrect. The formation of a glassy texture depends on rate of cooling, not on whether the cooling took place above or below water (see Chapter 1).

After further misleading statements, Robinson (p. 23) concludes that

> the circumstantial evidence in favour of the Atlantean theory continues to accumulate. Men of science are beginning to give more credence to the idea.

Belief in this kind of evidence would be restricted to the naive and uncritical. A final comment on Robinson: Atlantis is portrayed as a highly advanced civilization equal or superior to our own, according to Robinson. Atlanteans had aircraft, submarines, lasers, radio, television, photography, telephones, and elevators. Despite all this sophistication, Robinson reports that once the Atlanteans knew the end was near, they searched for suitable lands to colonize, and that (p. 62) "many expeditions went out for this purpose." This seems curious. Such an advanced civilization would be quite familiar with earth's geography, and what was and what was not suitable for colonization would be well known. They would not have had to find out.

## The Real Atlantis

Can the real Atlantis be found, given the fiction, fantasy, and speculation that encrust this legend? The problem is compounded by the passage of so many centuries both before Plato and since.

It does not seem reasonable, if Atlantis existed and died dramatically, that the sole source for the memory of such a cataclysm would be one man (either Plato or his ancester, Solon). John S. Bowman, in his *The Quest for Atlantis,* makes a useful point about this when he says (p. 22):

> Solon told the story to Critias' grandfather, who when he was about ninety years old told it to Critias, then a boy about ten. Are we to believe that the grandfather, through all those years, never mentioned Atlantis to any other person—a relative or friend? If he did, why did not even a rumor of it get passed on to other people, taken up by writers or poets, or mentioned somewhere? Much of the literature and writings of those years has been lost forever, but much has survived, in one form or another. Yet not a fragment, not a line, not a footnote or cross reference to this story of Solon's exists until Plato refers to it.

If it is reasonable to assume that Plato created the legend, we would admit that there may have been some historical basis for the idea. We have mentioned the dearth of corroboration for Plato's story in pre-Platonic times. However, L. Sprague de Camp, in his detailed analysis of the Atlantis theme, notes that one writer, Thucydides, had already described small-scale quakes and floods within the Mediterranean:

> At about the same time, while the earthquakes prevailed, the sea at Orobiai in Euboia receded from what was then the shoreline, and then coming on in a great wave overran a portion of the city. . . . In the neighborhood also of the island of Atalante, which lies off the coast of Opuntian Locris, there was a similar inundation, which carried away a part of the Athenian fort there, and wrecked one of two ships which had been drawn up on the shore.

"Now," says de Camp, "if any passage in pre-Platonic literature, still extant, gave Plato his idea for the sinking of Atlantis, this is it" (p. 236). De Camp does not conceal his attitude toward Plato, for a little later on he remarks (p. 254):

> If all Atlantists would read the rest of Plato's writings as well as his Atlantis story, and realize what a fertile myth-maker he was, they might be less cocksure about his reliability as a historian.

It is probable that the legend of Atlantis would never have gained significant fame, despite Plato, if it were not for an inescapable geologic fact: Pleistocene glaciation. It is a matter of fact—not theory or speculation—that

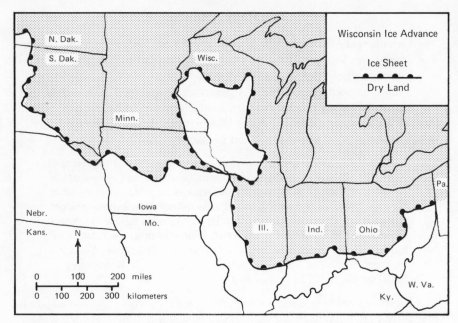

Fig. 10-5. Maximum continental ice advance during the Wisconsin lowered worldwide sea level. With ice retreat, sea level rose along world shorelines, creating flood legends.

worldwide temperatures declined for several tens of millions of years following the extinction of the dinosaurs, and that decline ushered in the Ice Ages between 1 and 2 million years ago. Thick sheets of ice—continental glaciers—formed in the northern hemisphere and spread south across Canada, parts of the United States, and western Europe. At least four major advances and retreats of the ice took place. Ice masses covered hundreds of thousands of square miles, and were two miles or more thick (Fig. 10-5). The withdrawal of enormous amounts of water from the world's oceans caused sea level to drop during each major ice advance. Geologists generally agree that during the last advance (the most extensive) sea level was lowered about 300 feet. The ice retreated and sea level rose, beginning about 12,000 years ago. It is still rising.

What this means is that any human constructions or other subaerial artifacts along world shorelines during the past 5000 years or more of human history have been flooded and submerged. The people have been forced to migrate to higher ground, and they have carried with them tales of floodwaters. For this reason, it is little wonder that most cultures have some stories or legends of great floods. They really happened.

# PART IV

# MONSTERS, STARS, AND CATASTROPHISTS

# Prologue

Man has come to rule this planet by virtue of his rationality, curiosity, and ingenuity, among other things. To a large extent we are incredulous and not a little surprised at our own accomplishments—from building the pyramids to getting to the moon. This attitude seems to reflect that somewhere within our collective consciousness, a small, somewhat awed child exists, observes, and wonders. The last four chapters of this book take up this notion.

The subject of monsters in Chapter 11 is one of endless fascination—looming shadowy creatures alien to ourselves, often repugnant, but always magnetic. To a small child, the world is full of giants, human and otherwise. This perspective seems to be carried over into the adult world where two "giants" still claim wide attention—the Loch Ness Monster and Bigfoot. Astrology is discussed in Chapter 12. Even in the 20th century, people have a need for explanations for their own behavior and a map of the future in black and white. They seek these in the movements and positions of dim and distant celestial worlds.

Finally, it is quite human to seek simple and satisfying answers to the riddles of a complex and obtuse Nature, who renders her art on a canvas fabricated through eons of time and overlayed with the hues of diverse and intermeshed materials and processes. The child often rejects that canvas printing and responds to catastrophism, which is embodied in Velikovsky's colliding worlds and Noah's ark, recounted in Chapters 13 and 14.

# CHAPTER 11

# Monsters

*A fearful creature . . . swift of foot and strong, whose breath was flame unquenchable.*

—PINDAR, FIFTH CENTURY B.C.

## The Fascination with Monsters

One needs no greater proof of the public's attraction to the monstrous or the bizarre than the success of "monster movies." More than 40 years after *King Kong* was made, people were still watching the giant ape climb the Empire State Building on the late movie. Dracula and Frankenstein, initially portrayed by Bela Lugosi and Boris Karloff during the 1930s, continue to stalk their victims in numerous remakes and variations on the original theme. More recently, dinosaur-age creatures such as Godzilla appear and destroy entire cities. Other sinister threats come from giant ants, spiders, scorpions, and strange beings from outer space. The reader will likely be familiar also with certain sea monsters, such as the great white shark of *Jaws*. The list of monsters is a long one.

It is little wonder then that popular debate continues on the purported existence of sea serpents, the Abominable Snowman or Yeti, the Loch Ness Monster, and other modern monster mysteries. As in the case of the other mysteries we have discussed, proper analysis of monster stories is hampered by abuses of logic and wishful interpretations of evidence.

## What Is a Monster?

To a large extent, what is monstrous is a matter of viewpoint. To a small kitten, a human being is a monster. The kitten is frightened, backing away and

205

hissing at an extended hand. Monsters are supposed to be frightening to someone, but this is hardly a good definition of a monster. More precisely, a monster is an animal or plant that departs greatly from the normal appearance of its kind.

The science of teratology is the systematic study of these deviations. Two main classes of monsters are recognized by teratologists:

1. Those with excessive or deficient growth in a single body (this category includes giants and dwarfs, also those born without eyes or brains)
2. Those with partial or complete doubling of the body on one of its axes (exemplified here are Siamese twins, and those born with two heads and one body, or two bodies and one head)

The causes of such abnormalities have been traced to defective genes and/or environmental conditions which have an adverse effect on the embryo in its very early stages of development. In humans, monsters may be produced as a result of exposure to X rays, diseases such as rubella, atomic radiation, and certain drugs such as the thalidomides.

In laboratory experiments with animals, teratologists have used both physical and chemical treatments on their subjects to produce monsters. In fact, most of the types of monsters recognized as such can be produced at will by using specific treatments to produce a certain effect on the developing embryo. While this may seem to be a rather cruel thing to do, it should be borne in mind that such experimentation, say, with rats, can lead to development of preventive measures to obviate heartbreaking abnormalities in human babies. Had teratology been sufficiently advanced, the rash of abnormalities resulting from pregnant women's use of thalidomide several years ago might have been avoided.

We now have some idea of what science thinks of as monsters. Applying this framework to the Loch Ness Monster, if it exists, and is indeed a holdover from the Mesozoic Age of Reptiles (which is not totally unthinkable), then it would not be a monster in the technical sense. It would be an ordinary member of the plesiosaur group, and not abnormal within that group. The Yeti of the Himalayas, if it is an abnormally large and hairy version of *Homo sapiens,* would be a monster. If, on the other hand, it is a member of a separate subspecies of man or ape sharing common characteristics, it would not be classified as a monster.

So, we return to an earlier statement that what is thought of as a monster depends on your viewpoint. We humans use the term very loosely to describe any creature of formidable and terrifying aspect. In the mythology of our race, monsters were thought to be the result of sexual intercourse between humans and devils, or between humans and various animals. These supposed creatures, however, had some element of humanness about them, perhaps making a caricature even more repelling. But also one might regard as monsters many creatures which are no more than ordinary members of their own species: the white shark, whales, and dinosaurs, to name a few (Fig. 11-1).

Fig. 11-1. The blue shark, often called a monster, is an ordinary member of its own species *(U.S. Navy)*.

## Some Human Giants

Excessive growth among humans is usually caused by a disturbance of the pituitary gland. The point at which such growth qualifies the individual as a "monster" is an arbitrary one. However, few would argue that a person exceeding eight feet in height departs radically from the norm of the species, and thus fulfills the technical definition of a monster.

The tallest man of recent record is Robert Wadlow (1918–1940), who stood 8 feet, 11 inches (Fig. 11-2). He weighed 419 pounds and his hands were 12 inches in length from wrist to fingertips. Wadlow toured with Ringling

Fig. 11-2. Robert Wadlow traveling with the Ringling Brothers circus. He was a giant and thus, technically, a monster (*Circus World Museum, Baraboo, Wisconsin*).

Brothers circus. There have been tall women, too. Jane Bunford, born in England in 1895, attained a height of 7 feet, 7 inches.

The most celebrated giant of the 18th century was Charles Byrne, born in 1761. He reached a height of 7 feet, 7 inches, and like Robert Wadlow, lived only to age 22. Byrne had a great dread of being put on exhibit in a museum after his death. According to one story, he arranged for friends to sink his coffin into deep water. A surgeon named John Hunter circumvented this plan by getting the burial party drunk and substituting rocks for Byrne's body. Whether this is true or not, Hunter did come into possession of the body and today the skeleton is on display at the Royal College of Surgeons in England.

Tales of giants abound in human history. This is one case where myth and legend have a basis in fact, as the preceding examples indicate. However, the size of legendary giants such as Goliath (more than 9 feet) and Og (about 11 feet) probably have been exaggerated. Also, some excavated bones have been thought to be those of a giant man as the result of mistaken identification of large animal bones as human. For example, discovery of enormous teeth belonging to the extinct ape *Gigantopithecus* gave rise to tales of giants 15 to 20 feet in height (Fig. 11-3). Such errors are understandable, but deliberate hoaxes have occurred also. Such was the case of the Cardiff Giant.

In 1869, a crew of men set about digging a water well on William Newell's farm near the town of Cardiff, New York. For a few hours the work had proceeded without incident when their shovels struck something large and unusual. Further digging revealed a gigantic petrified human figure 10½ feet tall (Fig. 11-4). Thus began one of the more famous hoaxes in American history. Newell began charging the curious for admission to see the giant, and eventually the Cardiff Giant was exhibited at carnivals and fairs around the country.

Although it fooled many people, some scientists disclosed that it was

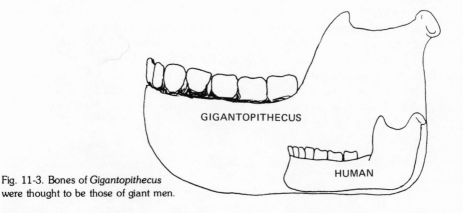

GIGANTOPITHECUS

HUMAN

Fig. 11-3. Bones of *Gigantopithecus* were thought to be those of giant men.

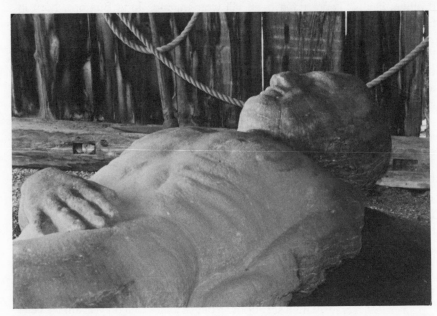

Fig. 11-4. The Cardiff Giant was carved from a block of gypsum. It was a deliberate hoax *(New York State Historical Association)*.

simply a large block of gypsum that had been carved into human form, and was of recent vintage. Gypsum is soft (it can be scratched with a fingernail) and easily carvable. The truth soon came out. A year earlier the statue had been carved by two artisans in Chicago at the behest of George Hull, who happened to be Newell's brother-in-law. The statue had been buried the previous autumn and allowed to "age" for a year before its "discovery." The two men made considerable money as a result of the hoax. Even after it was generally known to be a hoax, people continued to pay money to see it. The Cardiff Giant can still be seen at the Farmer's Museum in Cooperstown, New York.

Hoaxes notwithstanding, giants have walked the earth. But the actual truth of stories of legendary giants is obscured by time, and rational analysis of these mysteries is difficult. However, one type of giant is said to walk the earth today, living in remote areas and avoiding human contact. This is known as the Yeti, or Bigfoot.

## Bigfoot

Louise Baxter, who lives in Washougal, Washington, was returning home one day after driving her mother to the airport. Suspecting a flat tire, she stopped the car to investigate. She is quoted by Robert and Frances Guenette

*(Bigfoot: The Mysterious Monster,* p. 71) as follows:

> When I was checking the right front wheel, with my back to the woods, I could feel something staring at me. I had this feeling of being watched and I turned around. And that was when I saw Bigfoot. It was the biggest thing I had ever seen alive, 10 to 12 feet tall, with huge broad shoulders, a long torso. It was brown, had a big huge, hairy body. . . . He was like a man, more man-like than an animal, and he seemed intelligent. Mostly because of his eyes. They were enormous and they were glowing. . . .

It is from such eyewitness accounts as this that the mystery of Bigfoot was established. Reports of this shadowy creature describe a large, bipedal hominid (?) covered with short reddish-brown hair except for face, hands, and feet, walking erect, and standing anywhere from 7 to 12 feet tall. The creature is shy, elusive, and not usually hostile. Suggestions that eyewitnesses saw a bear, ape, or other animal are vigorously denied by the witnesses.

This creature, or a group of similar creatures, is known by many names. In the Himalayas the Sherpas speak of the Yeti, or Abominable Snowman (Fig. 11-5). The northwestern United States and nearby Canada is the stamping ground of Sasquatch, or Bigfoot.

The presumed existence of Bigfoot is based chiefly on the following: (1) eyewitness accounts, (2) tangible evidence, and (3) photographic evidence. Much of these data are summarized in the book by the Guenettes, from which we quoted the Louise Baxter testimony. This interesting book, although sympathetic to belief in Bigfoot's existence, was written with an obvious effort to be objective, as in the case where the authors report that a farm woman saw a "Bigfoot" which turned out to be her own cow. We will respond to the pertinent subject matter of this book under the three headings listed above.

Fig. 11-5. The theme of the Yeti or Abominable Snowman portrayed in postage stamps of Bhutan.

## Eyewitness Accounts

People say they have seen Bigfoot. An important question is, how good is the quality of this evidence? Two opposing viewpoints are expressed in the pages of the Guenettes' book. On the negative side, here is what Dr. T. D. Stewart has to say (pp. 62–63):

> It [eyewitness testimony] isn't very good because it can't be tested. . . .
> These people, they *want* to see something strange. They can imagine it.

As opposed to this, Dr. R. H. Rines maintains (pp. 23–24):

> Eyewitness testimony hangs people in a court of law. . . . So I don't
> understand how organized science can ignore it, can just dismiss reliable
> accounts by reliable people.

We find some fault with both these views. Dr. Stewart presumes to understand the psychological makeup of all the people who claim to have seen Bigfoot. They are all imagining something they want to see, he believes. We think such generalizations risky.

On the other hand, Dr. Rines points out that eyewitness testimony hangs people in a court of law. That is no doubt true in some cases, but only with corroboration and supportive evidence. Along the same line, a sharp attorney would tear apart the testimony of many eyewitnesses because of their excitement, brevity of the sighting, darkness, poor lighting, or other undermining circumstances.

Dr. Rines would also have science accept testimony as true simply because the eyewitnesses were "reliable." What is meant by reliable? The Guenettes refer to witnesses as "respectable," or youthful witnesses as clean-living people not on drugs or alcohol. There seems to be a rough equating here of reliability with integrity. The Guenettes ask (p. 103):

> Is it possible that so many people over a period of time could fabricate such
> stories and maintain them? Were they telling lies? Or were they telling the
> truth?

The implied (and very weak) reasoning here is:

1. Mr. Jones said he saw Bigfoot.
2. Mr. Jones is respectable and honest.
3. Therefore, Bigfoot must really exist.

To some extent, a parallel can be drawn here between UFO eyewitnesses and those who claim to have seen Bigfoot. Again we have things seen by honest witnesses, but perhaps misinterpreted. Often in the pro-Bigfoot literature we see photographs of eyewitnesses undergoing lie detector tests. This should not be construed as proof that Bigfoot actually exists, but only that the honest witness *thought* he saw something that *might* have been Bigfoot (or *might* have been a bear).

The pro-Bigfoot literature provides us with several eyewitness accounts. We assume they are the strongest and most convincing available from a spectrum of sightings. Yet even these "prime" cases are often vague and uninformative. For example, Jim Craig and Colleen McKay saw "something" near Portland, Oregon, in 1968. Despite observation for a period of five to ten minutes, their account boils down to the following facts as reported by the Guenettes: (1) it was big, twice as large as a normal person; (2) it was a dark, dark brown; (3) it was seated, legs extended and arms hanging down. This doesn't give science much to go on, nor does the sighting by Louise Baxter, cited earlier.

We do not question the integrity of any of these people, but could not Baxter's sighting be something of an exaggeration, at least as to size of the being she observed? How good is Baxter's judgment of the height of objects, especially under the unusual circumstances? A 12-foot monster would be three feet taller than the biblical Goliath. Such a monster could peer into second-story windows of an average house. It is difficult to imagine such a family of beings living undetected at the edge of civilized northwestern United States, with campers, hunters, tourists, and low-flying forest fire observers criss-crossing the region (Fig. 11-6). In addition, rewards of up to $100,000 have been offered for the capture of a live Yeti or Bigfoot. Despite this incentive, no one has yet been able to claim this reward.

On the other side of the coin, the more the eyewitness testimony piles up, the more difficult it becomes simply to dismiss the whole matter, even when the cranks and the hoaxers are taken into account. There is some strongly persuasive evidence noted by George Gill, who gathered as many eyewitness reports as he could find that included size of the creature, hair coloration, and footprint length. These data included sightings ranging from California to British Columbia. Here is what Gill has to say after he made some statistical inferences (personal communication; see also "Vancouver Population Clines . . ."):

> The average track length . . . varies gradually in a south–north direction from 15 inches average in California to 18½ inches in Canada. Corresponding to this is a gradual increase in average stature from 7 feet 4 inches in California to 8 feet 8 inches in western Canada. Of key interest here is the fact that this suggested increase in body size from south to north within the species range is exactly what one would predict in nature according to known zoological principles. Additionally, during my survey of the Sasquatch literature, I noticed that not only size but coat color varied in the same north–south manner, and the variation follows rather closely another zoological prinicple known as Gloger's Rule. It states that lighter coat colors within mammalian species tend to occur at the more northern latitudes. In fact among the California reports surveyed, 100% of the

Fig. 11-6. Signs such as this one on Route 20 in the state of Washington mark places where Bigfoot has been seen *(Jim Mueller)*.

Sasquatches were listed as brown or black, while among those from further north in Oregon, Washington and western Canada 80% were reported light gray, beige or white and only 20% brown or black.

As Gill points out, it would be impossible to fake such trends, and they are a major point in favor of the credibility of eyewitness reports.

## Tangible Evidence

Major evidence for the existence of Bigfoot is footprints found in mud and snow. These range in size up to 20 inches by 9 inches. Hundreds of plaster casts have been made by professional and amateur investigators alike. Although some of these prints doubtless have been faked, there are many that are difficult to explain.

The most common way to fake the footprints is to carve large feet out of a plank of wood and nail ordinary shoes or boots to them. Fake tracks can then be made in much the same manner as one uses snowshoes. Tracks that do not seem to have been fabricated in this manner are deep tracks (suggesting heavy weight) with a longer than normal stride, and sometimes seen to ascend unusually steep slopes. If these tracks are fakes, somebody went to a great deal of trouble, using techniques not readily discernible.

Other kinds of evidence include hair, fecal droppings, bones, and purported sounds made and recorded of the Yeti. The evidence from hair and fecal droppings is a bit shaky. In the first place, few samples of these materials alleged to have been produced by a Bigfoot have been submitted for analysis. Second, there is no basis of comparison. We can obtain samples of bear hair and droppings from such animals in captivity and compare them with samples obtained in the wilderness. Experts can identify human hair, even to its race. Animal hair is less easily identified. None of the hair thought to be from Bigfoot has been found to be human. If an analyst states that the origin of such hair is unknown, one need not infer, as some writers have, that it is from Bigfoot.

Scientists say: show us a body, or bones from such a body. The pro-Bigfoot investigators point to a body encased in ice (called the Minnesota Iceman) as perhaps the corpse of a young Bigfoot. This body is part of a traveling carnival. Apparently, no one has been permitted to thaw out this creature to find out exactly what it is. Seeking scientific evidence at a carnival exhibit is fraught with difficulties.

There is a lack of actual bones to examine. What bones are said to exist are very elusive. Some are said to be in a museum, but no one can find them. The Russians are said to have found two skulls in 1964 (Guenette and Guenette, *Bigfoot*), but no descriptions have been published. It has been suggested that Bigfoot buries his dead, hence there are no bones at the surface. In fact, one eyewitness, cited in *Bigfoot,* claimed to have observed a group of them burying one of their kind. This seems odd. Most Bigfoot investigators claim Bigfoot is not a tool user. How did they dig the hole? Why could not the eyewitness have led investigators back to the burial site?

We do agree that if Bigfoot exists, the bones would be rare. As Bigfoot hunter Peter Byrne rightly points out, terrestrial animal remains have little chance of preservation because of abundant scavengers and chemical decay.

The noises that Bigfoot makes have been heard and recorded, according to the Guenettes. These sounds are described by witnesses who both saw and heard the creature, and by those who only heard the sounds, but ascribed them to the monster. Here is an abbreviated list of the kinds of sounds:

1. Harsh grating moan
2. Half-bark, half-growl
3. Chattering
4. Strange whistling sounds
5. High-pitched scream
6. Half-laugh, half-language
7. Bird calls
8. "It sounds like animals, people, everything all combined . . . tapers off into a whistle."

What is a scientist, or anyone, supposed to make out of such a conglomerate of noises? The whistling and screaming are supposed to be the most common sound. But if all or most of these sounds come from the throat of one kind of creature, it has a remarkable set of vocal cords. Maybe so. However, no scientist would consider these data to be evidence supportive of any conclusion other than that the forest yields a variety of noises.

## Photographic Evidence

Photos and motion pictures have been taken. Scientists are usually suspicious of photographic evidence, especially when the photos are blurry and indistinct. Unfortunately, the photos we have seen are in this category. Some are so bad, in fact, that we had to study them for several minutes before making out what we were supposed to be seeing. These are the "good" photos. The Guenettes think the still photos are poor and inconclusive. We agree.

The Guenettes, however, believe the motion pictures taken by Roger Patterson in 1967 at Bluff Creek, California, constitute the strongest piece of evidence for the existence of the Yeti, or Bigfoot. These movies are now well known, having been seen by large television audiences. The question boils down to whether or not the large, hairy creature is real, or a man in a specially made-up suit. We have seen the film several times. The creature moves like a human, in our opinion, but that does not mean it is a fake. Primatologists disagree: some say it is not a hoax; others have "grave doubts" about the film's authenticity. Hollywood makeup men say it is not a suit, and that such suits are very expensive to make.

## Summary

We are not saying that Bigfoot exists or does not exist. We do not agree with the Guenettes that the evidence is necessarily additive. Fuzzy photos added to recorded noises from a forest do not make a Bigfoot. Several weak links do not make a strong chain. We also disagree with those who maintain that the scientific community airily dismisses the entire subject as "nonsense." If we can speak as members of that scientific community, hard evidence is important, and there is very little of it, if any. So much of the evidence boils down to unverifiable reports, some of which are no more than vague gossip, vague descriptions, and wishful thinking. However, there is sufficient evidence for us to believe these investigations should continue. Scientists are attracted to the mysterious and unknown as much as anybody.

## The Loch Ness Monster

Loch Ness is a long finger of serene, deep fresh water in Inverness, Scotland, 22½ miles long and about 1 mile wide. More than 1400 years ago a strange creature of gigantic proportions and a serpentine neck emerged from the depths of the Loch and made to attack a helpless swimmer. Nearby was a monk, Saint Columba, who shouted at the monster, "Think not to go further nor touch thou that man. Quick go back!" The creature at once backed off and sank beneath the surface, and the life of the swimmer was saved. This episode, apparently, was the beginning of the legend of the Loch Ness Monster.

The creature was seen again and again through the Middle Ages, and up to the present day. The question, does it exist? is answered in the affirmative by many. The follow-up question, what exactly is it? is still a matter of debate, conjecture, and continued observation and study of the surface of Loch Ness and its depths. The Loch holds its secrets well, by virtue of its great depth—950 feet in places—and the peat-stained waters, which drastically limit visibility for skin divers and underwater cameras alike.

## Evidence

The chief evidence for the existence of the Loch Ness Monster is of the same type as for Bigfoot; eyewitness accounts and photographs. Eyewitnesses generally agree on a reptilian creature with a long neck, small head, and large body. Often, one or more humps are seen protruding from the water, and the creature is fast-moving. Turbulence across the surface may mark its passage just below water level.

The eyewitness testimony is better in this case than the case of Bigfoot. For the latter, there are any number of causes for mistaken identification in sightings reported from all over the world. On the other hand, the Loch is a concentrated stretch of only 27 square miles of open water, familiar to the surrounding inhabitants who comprise the chief eyewitnesses. There is a severely restricted number of possibilities to explain what honest observers are seeing there. We do not believe these people on the Loch are seeing crocodiles, squids, or hippopotami. Nor do we think all of these sightings are floating logs or tree trunks.

The photographs are another matter and, as usual, cause problems. Some are fakes. Others are completely out of focus and obscure. The classic photo,

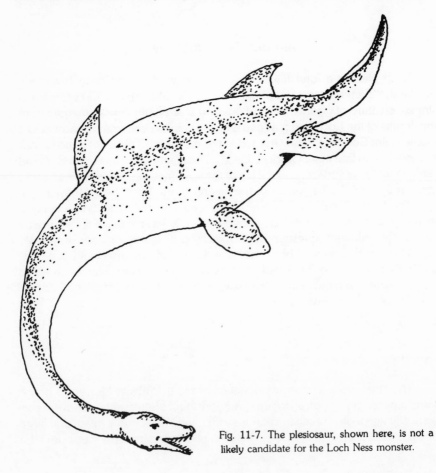

Fig. 11-7. The plesiosaur, shown here, is not a likely candidate for the Loch Ness monster.

taken in 1934 by the London surgeon Robert K. Wilson, is not conclusive either. We have examined this photograph on a number of occasions. We can only venture an opinion, but the waves do not seem right. There is no scale, but to us the "waves" look like small ripples one to two inches in height. If the height of the monster's head and neck is, say, six to eight feet above the water, which fits in with the proportions given by eyewitnesses, then the waves should be perhaps two to three feet high. Such waves ought to be breaking somewhere along their crests, but the photo does not show this. The waves (ripples) would be perfectly normal as shown if this were, as has been suggested, an otter's tail still in view at the moment of plunging into the water. An otter's tail is tapered, like the monster's neck in the photo. We think the photo is of an otter's tail, but pose this only as hypothesis based on the observations above.

In any event, we have eyewitness testimony suggesting that something is out of the ordinary. The latest photos, obtained in 1972 and 1975, are the best yet, especially one series of photos taken underwater which seem to depict a flipper. Sonar soundings have been made which suggest a large object or objects moving along which are probably not schools of fish. All of these items are provocative, but do not answer the question of what it is.

As in the case of Bigfoot, there is no solid physical evidence. No carcass has ever floated to the surface or washed up at the shore. No bones have been reported. Tasty chunks of "bait" lowered into the water have been ignored. Despite this lack of hard evidence, we agree that there must be something unusual in the Loch. Again, what is it?

## Portrait of a Plesiosaur

The most often voiced theory about the Loch Ness Monster is that it is a plesiosaur. The creature's neck and head do resemble what a plesiosaur ought to look like. The plesiosaur is one of several types of sea-going reptiles that lived during the Age of Reptiles and are now extinct. It was anywhere from 10 to 60 feet in length (Fig. 11-7). The theory is that a group of these creatures survived the general extinction of their kind at the end of the Cretaceous Period, about 70 million years ago, and somehow found their way into the Loch.

It is agreed by all that the creature cannot be a single entity because it has been reported in the Loch for a minimum of 1400 years. Thus, a herd of such animals has been theorized which, in order to keep surviving, would number at least 20 to 25 individuals. Supporters of the theory are quick to point to the remarkable discovery of the coelacanth, a fish thought to have been extinct for as long as the plesiosaur, and first described from fossilized remains in rock. Live coelacanths have now been discovered in African coastal waters. If it could happen to the coelacanth, it could happen to the plesiosaur. The point is well

taken, but simply indicates the possibility, not the likelihood, that the plesiosaur could survive. It seems to us that followers of the plesiosaur theory have not considered seriously enough what science knows about the plesiosaur.

The presumed habits of the monster are contradictory. It appears at the surface only rarely (one sighting for every 350 hours of observation, according to members of the 1972 and 1975 expeditions to Loch Ness) and it prefers dark, deep fresh water, lurking in underwater caverns at depths of up to 950 feet. The paleontological record of the plesiosaur indicates a shallow marine, air-breathing reptile. If there were plesiosaurs in the Loch, we would have a group of 20 or more individuals charging up and down the Loch close to the surface, chasing schools of salmon or other fish, and frequently breaking the surface to gulp air. The movement of the powerful, oarlike flippers as the plesiosaurs wheeled and dived to pursue their prey would create considerable churning action, as paleontologists have inferred from the bone structure. We would guess that, considering the confined surface area of the Loch, instead of one sighting for every 350 hours of observation, there would be a much higher frequency of sighting.

Suggestions that the plesiosaur could have adapted to fresh water and developed gills to remain under water are without scientific foundation. Adaptations of that magnitude would take millions of years in the course of organic evolution, and the Loch has only become free of ice since the recession of glaciers 10,000 to 12,000 years ago. Thus, the plesiosaur would have had to survive in the world's oceans for 70 million additional years before entering the Loch 10,000 years ago. Why should it force its way into a confined body of water alien to its existence when the oceans still remain much as they were as the natural home of the plesiosaurs of millions of years ago?

## Eels

Alternatively, we like the eel theory. There is a substantial eel population in the Loch. Those eels that did not make the journey to the Sargasso to spawn would grow larger, perhaps to impressive proportions. Based on eyewitness descriptions, silhouettes of the Loch Ness Monster are shown in Figure 11-8. These seem to confirm a more eellike form than any other beast, living or extinct. The eel also fits in better because it is not air breathing and is a lover of the dark. This would perhaps cut down sightings to the level observed in recent expeditions to the Loch. A final provocative item is the six-foot larval eel caught

Fig. 11-8. Silhouettes of sightings of the Loch Ness Monster show a humped creature. Is it a large eel or a plesiosaur? Or neither?

by the Galathea Deep Sea Expedition. As an adult, such an eel could equal or exceed the reported dimensions of the Loch Ness Monster.

## Wyoming's Little People

We have dealt exclusively with giants among the monsters in this chapter. We conclude with a brief look at another kind of monster known as Wyoming's Little People. It is a fine example of how a legend can have a basis in fact, although the legend does not convey the whole truth.

Among the Shoshone and Crow Indians of central Wyoming, tales are told of a race of miniature people of human appearance standing only inches high who inhabit the mountainous area, perhaps living in caves and generally avoiding their larger brethren. Such stories would be disregarded by a sensible person, relegating them to the same category as fairies and hobgoblins. Yet the stories persist, even among some of the local cowboys.

The mystery started to unravel in October 1932, when two prospectors, Cecil Main and Frank Carr, were poking around in the Pedro Mountains southwest of Casper, Wyoming. Using dynamite, they accidentally blasted open a sealed cave entrance. Gingerly, they moved into the cave passage

Fig. 11-9. "Pedro." While it is a monster, it can hardly be described as one of dread *(Ivan P. Goodman).*

which led to a cavern. There, seated on a rock ledge, was a small humanoid, mummified creature. In its sitting position, the small mummy measured only 6½ inches in height (Fig. 11-9).

Could this be one of the "Little People"? Scientists examined it in detail, taking both photographs and X rays. George Gill of the University of Wyoming agrees with other scientists that, despite its "old age" appearance, the mummy is that of an infant or fetus.

This mummy, called Pedro, is only one of several that have now been found, averaging 14–15 inches in height. Physicians and anthropologists diagnosed these infants as probably suffering from anencephaly at the time of their birth: an absence of part or all of the brain. This malady also imparts to the victim the appearance of a tiny old man rather than an infant.

It is speculated that a tribe with a high incidence of anencephaly might have left these infants in caves where they underwent mummification. Later tribes might have discovered them many years later, giving rise to the legend of a race of tiny people somewhere in the mountains.

# CHAPTER 12

# Astrology

*. . . in this age, astrology will regain its lost respectability.*
— JOAN QUIGLEY, ASTROLOGER, 1969

## The Daily Horoscope

Pick up a newspaper sometime and read your horoscope for the day—perhaps you already do this regularly. What does it tell you? Perhaps it will provide the same type of information we found in some typical examples selected at random:

> ARIES (March 21–April 19): This is a good day to spread ideas, but not to start ventures. Get rid of the useless, outworn. . . .

And in the same column: Geminis are urged to avoid flippancy and arrogance, Sagittarians should "be satisfied with a fair share of profits," and a Libra ought to "think in terms of success."

Most of this sounds like good advice applicable to anyone—to treat people kindly, avoid greed, and think positively. But should Aries avoid starting new ventures? Why? The astrologer would reply, "Because the positions of stars and planets and their relation to one another exert a certain set of influences that render some activities favorable in outcome and other activities unfavorable."

Orthodox science generally takes a hard line against astrology. It is thought to be a pseudoscience. Astronomers, the scientists most closely connected to the astrologers' domain, reject it as having no scientific basis. But regardless of what science thinks, millions of people follow astrology, with varying shades of belief in it. Many are total believers. Recently a newspaper inadvertently omitted its astrology column and the switchboard was besieged by irate callers, some of whom claimed that they were unable to plan their day unless they first read their horoscope.

Astrology is a multimillion-dollar business. The newsstands are sprinkled with magazines and books on astrology. There are numerous astrologers who make their living in private consulting, and there are some who operate astrology schools.

We are not astrologers, or even astronomers. There is little in astrology that impinges on our fields of geology and archaeology except the occasional discovery of artifacts that seem to have astrological symbolism. Be that as it may, we think much of the data of astrology are amenable to analysis using logic and reasoning in the evaluation of the evidence. We believe it is a worthwhile subject to include in this book because it has such widespread interest and is surrounded by unresolved mystery, as we hope to make clear.

Unfortunately, many scientists who condemn astrology out of hand know virtually nothing about its historical development, how it works in practice, or what it claims. Most scientists would be unable to cast a horoscope. Although our own knowledge of astrology is modest, we can at least cast a fairly accurate horoscope, the idea being that one should understand a subject in order to pass intelligent judgment on it.

We have heard scientists denounce astrologers because they allegedly attempt to foretell the future. Many astrologers would object to being put in the same class as fortune tellers. To them, the stars *"impel, but do not compel."* In any event, it seems to us that seeking out the future, being able to predict, is a primary objective in much of scientific research, such as studies aimed at predicting earthquakes, drought, floods, and famine. Furthermore, it would be difficult to convince a severely sunburned person that astral bodies such as the sun do not exert *some* influence over human affairs. On the other hand, it is hard to believe that the planet Jupiter's position will somehow be important in the financial transactions of a man's ice cream parlor.

To gain a preliminary understanding of astrology, let us review the historical roots of astrology and its subsequent growth.

## Historical Background

Human interest in the stars and their meaning has existed for thousands of years. Exactly when systematic observations began is difficult to say, but it is generally believed that astrology became established in Mesopotamia about 3000 B.C., perhaps even earlier. The origins and reasons for continuous observations are complex. We think it may have come about with the help of the following catalysts:

1. The desire to know and predict the future, especially among leaders and kings of ancient peoples
2. Recognition by the earliest observers of the repetition of celestial phenomena
3. The religious concept that the stars and planets were gods, each with special attributes for good and evil
4. The inference that people born at the time of the ascending of a particular star (god) might possess specific personality traits, strengths, and weaknesses imposed by that god.

We know that ancient kings as long ago as 5000 B.C. (and probably before) wished to know their fate, and recruited wise men and priests to examine animal entrails and use other commonplace means of divination. In comparison, astrology might have seemed to have a bit more verisimilitude. In any case, astronomical records were kept. Some of these records are still extant, and date back to 747 B.C. The data include movements of the sun and some planets, lunar phases, and eclipses. However astrology may be viewed today, it was the earliest exact science in human history.

The knowledge and practice of astrology either arose independently or spread by diffusion from Mesopotamia to many other places—Egypt, Greece, and as far away as India and China.

Egyptian observation of the heavens must have long endured, as can be seen by the orientation of all the major pyramids (except the Step Pyramid) to the cardinal points of the compass. The Egyptians constructed a calendar of 12 months, each of 30 days. It was based on the risings of 36 conspicuous stars, each separated by 10 days. Each of these bright stars was considered a god who ruled for 10 days. Later writers referred to these stars as *decans*. Astrologers today divide each sign of the zodiac into a first, second, and third *decanate* of 10 days each. They claim that the most intense influence of a sign occurs during the first decanate.

The Greeks were latecomers in astrology, getting involved about 400 B.C., but as might be expected, they had a strong influence. The zodiac, consisting of the 12 familiar signs, was a product of the Greeks, developed over several centuries. Our word *horoscope* comes from the Greek *horoskopos*, which means "decan star." However, it has come to refer to the map of the heavens at the time of one's birth. Each such horoscope is unique because the heavens, in aggregate, will not return to the same position for another 25,000 years. The Greeks followed the tradition of naming stars after gods. In addition, the Greeks invented the world's oldest scientific instrument, known as the astrolabe, used to measure the altitude of celestial bodies, from which time and latitude can be determined.

At the time the Romans came into contact with astrology, they rejected it. To them it was another religion and they had their own religion. In fact, astrologers were persecuted. This attitude changed by the time of Augustus Caesar, and astrology came to be accepted. The change may have been facilitated by the spread of the personal horoscope. Until about 200 B.C., a horoscope was something reserved for kings and other important people. After that time, the casting of horoscopes spread to the common people. There were probably too many individuals involved in astrology for the practical-minded Romans to bother maintaining a ban against it.

During the Middle Ages, official Christianity was anti-astrology, but it was of little use in restricting its growth. Astrology persisted, and even influenced the naming of the days of the week and the date for the Christmas celebration. Astrology was not confined to the uneducated and superstitious. Such scientific luminaries as Isaac Newton, Johannes Kepler, and Tycho Brahe served as astrologers and made part of their living casting horoscopes for wealthy people (whether these men really believed in what they were doing is another matter). Astrology even crept into the works of William Shakespeare, the English playwright, as when Cassius says to Brutus that "it is not in our stars but in ourselves that we are underlings." Yes, it would appear that astrology has been very popular since the Middle Ages.

In modern times astrology has burgeoned. There are now specialized horoscopes for love and marriage, careers and jobs, the stock market, and even for people's pet cats and dogs. Business executives, film stars, and political figures can be numbered among the followers of astrology. The German dictator Adolf Hitler had an astrologer, and it is interesting to speculate that the course of World War II may have been influenced by the judgments of an astrologer.

## How Astrology Works

**Major Influences.** The sun is the single most important influence in determining basic character, according to its position in the zodiac which divides the sky into 12 equal parts. Thus, if the sun is in the constellation Virgo at the time of one's birth, that makes one a Virgo native and in supposed possession of a certain set of traits. However, the position of other celestial bodies can serve to modify these basic traits in an individual.

The position of the moon is the second most important element in a person's horoscope and influences the kind of personality that person would have. To give an example, if a person were born with both the sun and moon in Virgo, this would tend to reinforce Virgo traits, such as a generally unprepossessing nature. But if the moon were in Leo, the individual might tend to be more assertive and dramatic.

One of the reasons an astrologer needs to know the time (even to the minute) and place of birth is to determine the rising ascendant. This is the constellation or sign on the eastern horizon, a different one rising every two hours. Astrologers are a bit vague about the significance of this aspect of astrology, but it seems to represent the physical self and represents similar traits of the respective sun sign.

**Planets.** In working out an individual's horoscope, planetary positions are also determined, that is, the sign in which each planet is situated at the time of birth. Each sign has a broad area of significance. For example, the sign of Cancer is associated with the domestic—with home and family. Table 12-1 lists the signs with brief characterizing descriptions. Each planet also has its own characteristics, shown in Table 12-2. As each planet moves into a sign, its special influence will be exerted on the area of life symbolized by that sign. For example, Jupiter in Taurus would be favorable for money matters, while the malefic Saturn's presence might indicate financial disaster. Venus in Cancer would suggest a serene marital situation, but Uranus in the same sign might favor divorce or family fights.

**Aspects.** Of importance, too, are the aspects, or angular relations of the planets to one another. These relations can be boiled down to conjunctions, trines, squares, and oppositions. Two planets are considered in conjunction if they occupy the same sign. In a horoscope, this may be good or bad, but it is mostly good. Trines occur between planets 60° apart and these are considered harmonious and lucky (120° is also good). Squares (90°) or oppositions (180°) generally signify obstacles and anxiety, but may be useful to some people who need challenges to bring out their character and skills.

**Table 12-1. Some General Characteristics Associated
with Signs of the Zodiac**

| Sign | Dates | House[a] | Significance |
|------|-------|----------|--------------|
| Aries | March 20–April 20 | First | Physical body |
| Taurus | April 20–May 21 | Second | Money and possessions |
| Gemini | May 21–June 21 | Third | Communications, environment |
| Cancer | June 21–July 23 | Fourth | Home, parents |
| Leo | July 23–August 23 | Fifth | Children, amusements, recreation |
| Virgo | August 23–September 23 | Sixth | Health, service |
| Libra | September 23–October 23 | Seventh | Partners, unity with others |
| Scorpio | October 23–November 22 | Eighth | Sex expression, self-sacrifice |
| Sagittarius | November 22–December 22 | Ninth | Travel, new horizons |
| Capricorn | December 22–January 19 | Tenth | Career, social status |
| Aquarius | January 19–February 19 | Eleventh | Friends, hopes, and wishes |
| Pisces | February 19–March 20 | Twelfth | Self-undoing, confinement, emotion |

[a]Houses equate roughly with the 12 zodiacal signs.

## Table 12-2. Characteristics Generally Associated
## with Each of the Planets

| Planet | Characteristics |
| --- | --- |
| Mercury | Governs understanding and sight; rules arms and legs, tongue |
| Venus | Harmony, friendship, love, good luck |
| Mars | Aggression, energy, action |
| Jupiter | Good luck, dignified, confident, jovial, compassionate |
| Saturn | Malefic, adverse, cold, conservative, serious obstacles |
| Uranus | Impulsive, explosive, revolutionary, alien |
| Neptune | Nebulous, emotional, psychic, inspirational |

Astrologers also note whether a sign is an earth, air, fire, or water sign, following the ancient notion of four basic elements. In general, two of the same elements (say, water signs Pisces and Scorpio) should get along well with each other. Water and earth signs are compatible. So are fire and air. But fire and water may not mix well. Signs are also labeled as cardinal (spontaneous), mutable (changeable), and fixed (deliberate, stubborn). Each sign is "ruled" by a particular planet.

When all of the factors we have mentioned are taken into account, the astrologer weighs and assesses them and produces a portrait of the individual's character, personality, strengths, weaknesses, and prospects. In so doing, some "psychic" astrologers might rely on hunches and intuition to aid their interpretation while others might be more inclined to use a computer.

To get a glimpse of future influences an astrologer will use the *ephemeris,* a list that gives planetary position and other astronomical data for the coming year(s). Thus, a client may be advised to avoid travel next March because Saturn, an adverse planet, will enter the ninth house (houses are roughly equivalent to the zodiacal subdivisions). The ninth house is the house of travel.

This, then, is the general picture of astrology. Astrologers claim that their science is the result of experience based on thousands of years of observation of people, their behavior, and the correlation of these and the events surrounding their lives with celestial data. The precise reasons why this should be are more obscure, but we will explore some of these possibilities later in the chapter. Before doing so, let us consider the objections raised by those who maintain that astrology is a pseudoscience.

## Objections to Astrology

**The Daily Newspaper.** Although we opened this chapter with a brief discussion of the daily horoscope seen in so many newspapers, we do not wish to give the impression that it is a serious objection because even the astrologers think it is a joke. The daily horoscope is based chiefly on the sun sign, and it is incomplete and so generalized as to be worthless.

**Moment of Birth.** This is a more serious objection. The astrologer says he must know the *exact* moment of birth in order to carry out his task thoroughly. Why? What is so sacred about this exact moment? The process of birth can take several hours. What is meant by the "exact moment"? The expulsion of the child is not accomplished in the blink of an eye. Even if we have a rapid birth narrowed down to short minutes, then what in the total condition of the newborn has altered between the moment prior to expulsion and the snipping of the umbilical cord?

We see little change except that the child begins to breathe, cry, and assume an independent existence. The genetic equipment of the child was already long preordained. The astrologer speaks vaguely of "cosmic forces" that affect the moment of birth, a sort of indelible marking of the child to produce a fateful proclivity toward certain traits that, apparently, did not exist even minutes prior to birth. If these cosmic forces are so influential, why are they not also efficacious during gestation? It seems to make as much, if not more, sense to pick the moment of conception as the critical time.

There is no scientific evidence that we are aware of to support the proposition that the moment of birth (except possibly a traumatic birth) has anything to do with the child's later personality, character, profession, or other life activities.

**The Rising Ascendant.** The rising ascendant can change in a matter of minutes during the process of birth. This is why, as we have noted, astrologers like to know the exact minute of birth. What difference does this make? Here is what Joseph Goodavage says in *Write Your Own Horoscope* (p. 137):

> The man or woman with a Sagittarius Ascendant, for example, loves outdoor life, travel, horses and sports in general. . . . Scorpio rising, on the other hand, is pretty much the opposite.

We are to conclude from this that an individual born at 4:30 A.M. on January 15 (Scorpio still ascending) will dislike horses, or at least be indifferent toward them, but if he or she is born five minutes later, at 4:35 A.M. (Sagittarius ascending), he or she will love horses. No concrete evidence is presented to support this claim.

**Time for Observation.** Astrologers say that much of the data of astrology is the result of thousands of years of celestial observation. Certainly, to determine, correlate, and verify the significance of the many intricate and complex stellar and planetary positions that form the foundation of astrology and relate them to human activity would require a very long time. Because slow-moving planets take many years, even generations, to move from one sign to another, we would have to predicate several tens of thousands of years of observations, extending back into the Stone Age.

Systematized and recorded celestial observations extending over many generations would require, in our judgment, a stable society with writing and communicative ability other than word of mouth. Therefore, what constitutes modern astrology has been brought along without sufficient time to have fully assessed the impact of the rarer stellar events such as eclipses and the passage of slow-moving planets from one sign to another. The significance of these events would have had to be based on inference, hunches, or simply guesswork.

**The Outer Planets.** Along with the ideas in the preceding section, it should be noted that Uranus was only discovered in 1781, Neptune in 1846, and Pluto in 1930. Horoscopes cast prior to 1781 without taking into account three planets should certainly have been incomplete and even misleading. Would not a few sharp astrologers before 1781 have suspected that many of their pronouncements were somehow not hitting the mark, and inferred some unknown planets out there? We have not heard of astrologers even voicing this possibility.

Likewise, it is difficult to see how, in the relatively short time since 1781, the influence of Uranus and Neptune could be evaluated and related to human behavior. Although Pluto, being small and distant, is thought to have only a minuscule effect on a horoscope, Uranus and Neptune—giant planets—cannot so be ignored.

**Ability to Predict.** Astrologers do not claim that there is an inevitability to future events, only that a future occurrence has high or low probability due to astral forces. Nevertheless, astrologers do make predictions, and the credibility of astrology to a large extent rests on success in this area.

When a prediction succeeds, it gains wide publicity. For example, Jeane Dixon foresaw President John Kennedy in danger in 1963 should he go to Dallas. Following the assassination of the president, Miss Dixon was widely acclaimed for her prediction. However, her prophecy would have been fulfilled, although less spectacularly, had Kennedy suffered some minor injury such as a broken leg in alighting from his plane. Apart from the above well-known example, we wonder what the "track record" is of all predictions by Miss Dixon and other astrologers. One seldom hears of the predictions that are wide of the mark. If astrologers are in a better predictive position than ordinary mortals, it would seem logical that the astrologers should be making considerable money on the stock market, at the horse races, or in lotteries. We know of no astrologer who has made a million doing this. However, many make a decent living writing books, articles, or newspaper columns.

Every year, a book usually appears in the paperback market that offers predictions for the upcoming year by a number of well-known psychics, clairvoyants, and other mentalists. One or two astrologers are usually included. Here is an example drawn from Warren Smith's *Predictions for 1975* (p. 201) in which scientific astrologer Cora Sitrusis reveals:

> December 1975 brings a more relaxed attitude, and a general feeling of well-being throughout the nation. The year 1975 appears to come to a conclusion with good unions and agreements that give cause for the holiday cheer. I do not find any reason for celebrating less than we have in the past years, as it is a more peaceful and serene time with sincerity felt by all.

We do not see how this qualifies as prediction, astrological or otherwise. One need not pore over horoscopes to decide that December, a holiday period, will be a time for "a general feeling of well-being" and so on. Most of the other predictions in this book are equally vague, yet the book cover announces that these are *astonishing forecasts*. It would be instructive if astrologers would publish a book each year listing specifically all forecasts from the previous year, working out the percentage of forecasts that were correct.

**Horoscopes for Inanimate Objects.** Often astrologers look for evidence among the planets and stars concerning future prospects of ships, planes, nations, and elections. This is puzzling to us. We can conceive of living organisms responding in some way to electromagnetic or other radiation so as to lend some credence to astrology. But a ship is a large piece of hollow metal, in essence. By what means can it be caused by the stars either to sink or to float? This makes as much sense as casting a horoscope for a can of vegetables.

## A Good Word about Astrology

We have heard it said that the stars and planets are too far away to have any effect on human behavior. We did not include this argument in the previous section because we believe, guardedly, that the argument is not altogether valid. We need not dwell at length on the more obvious interaction between the earth and celestial bodies. There are of course gravitational attraction and tidal effects. There is radiation from the sun in the form of heat and light affecting all living organisms, directly or indirectly. The sun exerts a strong influence on planetary weather, and this in turn can affect moods and activities. So, to this extent, there is an "astrology." Some would prefer to call it "cosmobiology."

Apart from the obvious relations we have been talking about, there are more subtle, invisible, but dynamic forces around us. The earth and its inhabitants are suffused and bathed in a variety of emissions at all times, which vary in intensity, variety, and source. These emissions or radiations are not just from the sun, which is an active cauldron of various emissions. They also come from the moon and other planets. Jupiter, for example, spits out bursts of low-frequency radiations as well as thermal and nonthermal radiation.

From beyond the solar system, radio noise pours in from the center of our galaxy. It is also reaching us from distant nebulae and galaxies. The spiral galaxy M87, in Virgo, throws out 1000 times more radio-frequency energy than does our own galaxy. These facts became known with the advent of radio astronomy in 1931, and this science has advanced dramatically since that time.

What has this to do with astrology? Maybe nothing. But it does demonstrate that energy can be received here on earth even from very distant galaxies. This cosmic energy may have effects on us we are not aware of. This possibility should not be dogmatically rejected until we know more.

There are certain observations from radio astronomy that, in a way, smack of astrology. For example, here is a quote from the *Encyclopaedia Britannica* in the section on radio astronomy:

> As the sun passes in front of this strong radio-frequency source [the Crab Nebula], the latter's signal shows irregular fluctuations.

This might be restated by an astrologer casting a horoscope: "With the sun in Taurus, you can expect certain changes to take place. . . ." It is speculation, but *if* the Crab Nebula's signal had a subtle effect on some individuals, the "irregular fluctuations" might cause some kind of behavioral response in those individuals. Such a notion is indeed fantastic, but is it more so than splitting and joining atoms, television, or trips to the moon? (We do not deny that ancient astronauts could have visited the earth, but only point out that the evidence to support this notion is insufficient. In the case of the Crab Nebula, the data of science itself is not forthcoming. This is as opposed to, for example, the use of a balloon to guide the drawings on the Nazca Plain in Peru, where a plausible alternative is available.)

## Astrology, Documentation, and Experimentation

Scientists might well be suspicious of data advanced by the astrologers to support their claims. The documentation is often nonexistent, or meager. For example, Joseph Goodavage relates this interesting story about astral twins (those born with identical horoscopes) in *Write Your Own Horoscope* (p. 27):

> A few years ago, two unrelated women met for the first time in a hospital room in Hackensack, New Jersey. It turned out that they had been born on the same date . . . both had the same first name—Edna. Each had come to the hospital to give birth to a first child. The babies were born at the same

hour, had weighed the same and had been given the same name—Patricia Edna.

The women's husbands also had identical first names—Harold. Each Harold was in the same business and owned the same make, model and color car. . . . Their husbands were of the same religion—a different one from that of the wives, which was also the same. . . . Each family owned a dog named Spot of the same mixed breed, size and age. . . .

We would agree that the parallelisms are quite remarkable. It is difficult to evaluate this evidence. The year is not given, the name of the hospital is omitted. A scientist investigating this type of phenomenon would like to see the birth certificates of these two ladies and otherwise verify the amazing similarities mentioned by Goodavage. Perhaps this was done, but Goodavage does not give his source for this case. There is not necessarily a suspicion that Goodavage is fabricating or exaggerating, but such an important chunk of astrological evidence ought to be documented. Unfortunately, it is not.

In contrast, biologist Lyall Watson, in his far-ranging book *Supernature,* dealing with unexplained phenomena, provides us with 347 references, mostly to scientific journals, to buttress and document his account of astrology and other topics. Dr. Watson thus provides his readers with the option of checking out his statements, an option denied to the readers of most astrology books. Here is how Watson reports on the investigations of Gauquelin, a French investigator (p. 49):

> In Europe all local authorities record the exact moment of birth in official registers, so Guaquelin was able to collect this information and match it with the positions of planets computed from astronomical tables. (119)*
> He selected 576 members of the French Academy of Medicine and found, to his astonishment, that an unusually large number of them were born when Mars and Saturn had just risen or reached their highest point in the sky. To check these findings, he took another sample of 508 famous physicians and got the same results. (120)* There was a strong statistical correlation between the rise of these two planets at a child's moment of birth and his future success as a doctor.

Even a quick reading of Watson's discussion of astrology, of which Gauquelin's research is only a part, broadly suggests that astrology is a field amenable to scientific investigation. Under stringent experimental conditions and standards, many of the claims of astrology could be either dismissed or verified. We see no reason why this should not be a valid arena for scientific

*The numbers in parentheses refer to the papers published by Gauquelin presenting results of his research. His results and conclusions, by the way, have been challenged and are the subject of debate among interested scientists.

investigation. Relatively little has been done. Most scientists who would be so inclined seem embarrassed to try. This is unfortunate, because it is only through the application of scientific method that the claims of astrology can be laid to rest or pursued further. In general, scientists perceive areas such as astrology, in contrast to the study of unidentified flying objects, as outside the domain of legitimate scientific inquiry.

Scientists lament, in our presumably enlightened age, that so many people continue to believe in astrology. Perhaps science itself is partly to blame. Scientists have greeted astrology, generally, with stony silence. The nonscientists are left with no guidelines except those placed there by the astrologers.

In 1975, a group of philosophers and scientists attempted to correct this situation. A statement appeared in *The Humanist* magazine (Sept.–Oct. 1975), signed by 186 leading astronomers and other scientists, which said in effect that astrology was a lot of bunk. Although it was unintended, this statement (which received fairly wide publicity) conveyed the impression of a fallacious argument, appeal to authority—in effect, that persons versed in astronomy or other sciences should be believed when they denounce astrology. No evidence is presented to show that they all are experts in astrology. Undeterred, this group formed a Committee for the Scientific Investigation of Claims of the Paranormal and began publication of *The Zetetic,* a forum for the discussion of not only astrology but the entire range of the paranormal, including most of the subjects in this book.

In its second issue, *The Zetetic* published an article by John D. McGervey, a physicist, entitled "A Statistical Test of Sun-Sign Astrology" (pp. 49–54). McGervey took the birthdates of 16,634 scientists and 6,475 politicians and

#### Table 12-3. Number of Births by Astrological Sign of Scientists and Politicians[a]

| Sign | Dates (inclusive) | Scientists | Politicians |
|------|-------------------|------------|-------------|
| Capricorn | December 24–January 19 | 1241 | 462 |
| Aquarius | January 23–February 18 | 1217 | 445 |
| Pisces | February 21–March 19 | 1173 | 460 |
| Aries | March 23–April 18 | 1160 | 432 |
| Tarus | April 23–May 19 | 1185 | 471 |
| Gemini | May 24–June 19 | 1153 | 471 |
| Cancer | June 24–July 20 | 1245 | 486 |
| Leo | July 25–August 20 | 1263 | 504 |
| Virgo | August 25–September 20 | 1292 | 497 |
| Libra | September 25–October 21 | 1267 | 523 |
| Scorpio | October 25–November 20 | 1246 | 488 |
| Sagittarius | November 24–December 20 | 1202 | 453 |

[a]There is no correlation (table courtesy of *The Zetetic*).

related them to the sun signs. There was no correlation between sun sign and a person's potential of becoming a scientist or politician (McGervey's results are shown in Table 12-3). Regardless of the results or the conclusions one may draw from them, it is heartening to see the beginning of scientific testing of the tenets of astrology, and we hope this will continue.

Finally, we might mention that we have made a few cursory investigations. The claim is sometimes made by astrologers that the time of the full moon is associated with sharp increases in crime. We investigated hospital and police records over an eight-month period for rapes, homicides, and admission to hospital emergency rooms in Buffalo, New York. There was no correlation at all between these incidents and the cycle of the full moon. However, it would be premature to publish these results with strong assertions until further investigation has been carried out.

# CHAPTER 13

# The Colliding Worlds of Velikovsky

*. . . it should be remembered that a law is but a deduction from experience and experiment, and therefore laws must conform with historical facts, not facts with laws.*

— IMMANUEL VELIKOVSKY

## Background

In 1950, Immanuel Velikovsky's controversial book *Worlds in Collision* appeared and became a best seller. This book offered a new and dramatic version of the earth, its people, and its history based on cosmic events, the most vivid of these being a near collision of Venus with earth about 3500 years ago.

Velikovsky's evidence for these events was drawn from a variety of sources: myths and legends; the Bible; ancient texts; and the data of astronomy, geology, and other sciences. His was an interdisciplinary approach, and he believed the diverse evidence "hung together" to form a unified, unique theory to explain history. Velikovsky's book is rich in footnoted references to the sources he uses to buttress his claims.

By invading the domain of science and challenging many of its accepted ideas and principles, Velikovsky soon came under fire from much of the scientific community. He was denounced as a fraud, a charlatan, and a quack. It is said that some scientists attempted to block publication of his book. He was considered unqualified to write such a book because his own training was in medicine and psychoanalysis. Many of the scientists who rejected Velikovsky's theories actually had not read his book. Even fewer have read, in ensuing years, his other books, *Earth in Upheaval, Ages in Chaos,* and *Peoples of the Sea,* which extended his views.

*Worlds in Collision* continued to sell, and 27 years after its original publication, the cover of the paperback edition (1977) proclaimed:

> A book of cosmic excitement! The international best seller that convulsed the scientific community . . . described primeval chaos when the heavens rained fire, continents writhed and shattered apart, and most of mankind was destroyed . . . and predicted with astonishing accuracy what space probes and moon landings would discover!

In this chapter, we wish to examine some of these claims. We cannot look at all of the evidence Velikovsky presents because some of it lies beyond our own ability to judge fairly. However, we can use elementary logic in assessing some of the evidence for his theories.

As a first step, let us describe more fully the major contentions made by Velikovsky. Before doing so, we should point out that Dr. Velikovsky is a *catastrophist* as opposed to a *uniformitarian.* The catastrophist believes that major changes of the earth are brought about by abrupt and violent cataclysms. The uniformitarians claim that major changes, generally, have taken place slowly, little by little over a long period of time (even millions of years) and by the same processes operative today as long ago.

## Velikovsky's Theory

As we understand it, Velikovsky visualizes the planet Venus being thrown off as a blob of gaseous matter from the planet Jupiter within historic times, and entering into a strongly elliptical orbit around the sun. Velikovsky refers to Venus as a "comet" because it had a long tail.

The comet Venus made a close approach to earth about 3500 years ago, coinciding with the exodus of the Israelites from Egypt. The earth entered the tail of the comet with calamitous results. Here is how it is described by the editors of *Pensée* in the preface to *Velikovsky Reconsidered:*

> In a few awful moments, civilizations collapsed. Species were exterminated in continental sweeps of mud, rock, and sea. Tidal waves crushed even the largest beasts, tossing their bones into tangled heaps in the valleys and rock fissures, preserved beneath mountains of sediment. The mammoths of Siberia were instantly frozen and buried.

It was also at this time that the plagues visited Egypt, brought about by the comet, and the pharaoh's army was drowned in the Red Sea after tidal attractions caused the water to recede and then return. Darkness enveloped the earth.

Following this catastrophe, the comet Venus kept its distance for 52 years before returning for a second encounter. This was during the time of Joshua, and another ruinous episode took place. The earth's axis tilted and the sun "stood still" amid the general destruction.

Centuries passed. Venus disturbed the planet Mars, forcing it into a new path so that now Mars endangered the earth. Again, from the preface of *Velikovsky Reconsidered,* here is how the editors of *Pensée* visualized the scene about 747 B.C.:

> As Mars drew near, the Earth reeled on its hinges. West of Jerusalem, half a mountain split off and fell eastward; flaming seraphim leaped skyward. Men were tossed into streets filled with debris and mutilated bodies. Buildings crumbled and the Earth opened up.

Mars came into close contact with the earth on two more occasions, in 721 B.C. and 687 B.C. These were apparently less destructive encounters except that, in the 687 B.C. meeting, 185,000 men of Sennacherib's army encamped outside Jerusalem were somehow burned to death and the earth tilted 10° on its axis. This is how Mars earned its reputation as the planet of war. After this, both Venus and Mars settled down into the nearly circular orbits around the sun where they are found today (Fig. 13-1).

## Critique of Velikovsky's Historical Evidence

**Nature of the Evidence.** The bulk of the evidence Velikovsky uses to support his theories comes from the almost limitless reservoir of ancient writings. We expect that much ancient writing refers to actual events, but one must be careful in using this type of evidence for the following reasons:

1. We are dealing with more than one writer. Thus, there will be an uneven reliability in terms of memory, honesty, imagination, and poetic license.
2. Much reporting was drawn from word of mouth accounts years, or even several decades after the actual event(s). This constitutes hearsay evidence.
3. Many references are couched in symbolism or mysticism with ambiguous meanings.
4. Many references have a religious import, with oblique meanings, perhaps understood readily by the people of the time but easily misinterpreted by the 20th-century mind.
5. Many natural events such as earthquakes and volcanic eruptions might be perceived by many ancients as supernatural, and from their provincial viewpoint, affecting the entire world.

Fig. 13-1. The planet Venus. Its composition as determined by space probes does not support Velikovsky's theories *(NASA)*.

6. Finally, there is the factor of the modern investigator, in this case Dr. Velikovsky, who must, in the face of the obstacles to truth we have just mentioned, *select* and *interpret* from this large body of literature. This entails the risk of unconsciously selecting only supportive testimony while being, shall we say, color-blind to that evidence which is non-supportive of a theory.

We think the difficulties enumerated above are real and must be guarded against. We do not doubt that Dr. Velikovsky was mindful of these problems as he progressed in his research, but our reading of his works leaves much of his evidence open to question. Let us examine some specific examples under separate categories.

**Unreliable Sources.** Pliny and Plato are examples of unreliable sources. Pliny died in 79 A.D. during the eruption of Vesuvius. He had been an avid and indiscriminate collector of miscellaneous information which he gathered into several volumes entitled *Historia Naturalis*. This work contains many errors and naive assumptions (such as the idea that soaking a diamond in goat's blood will rob the diamond of its hardness).

There are at least 16 references to Pliny in *Worlds in Collision*. Hearsay evidence is also involved, as when Velikovsky states (p. 293):

> In a no longer extant passage of Pliny there was something said about comets being produced by planets.

It is difficult to regard this as solid evidence that Jupiter produced Venus as a comet, which is what Velikovsky is getting at.

Plato is referred to several times, especially with regard to the lost continent of Atlantis. Although Velikovsky has his doubts about the accuracy of some of Plato's statements, he does not hesitate to link the alleged calamity of Atlantis to the catastrophic approach of the comet Venus. Velikovsky also seems to think that Plato *revived* the legend of Atlantis (p. 156) rather than *created* the legend, as seems more probable (see Chapter 10).

**Exaggerated Evidence.** Some of the fragments Velikovsky culls from early writings in support of catastrophic flooding seem so preposterous that we wonder why he includes them, and thus detract from the veracity of his whole theory. Here is an example (p. 87):

> The Midrashim contain the following description: "the waters were piled up to the height of sixteen hundred miles, and they could be seen by all the nations of the earth." The figure in this sentence intends to say that the heap of water was tremendous.

Note that Velikovsky chooses to place his own modifications on this direct quote. He has done this elsewhere as in changing Plato's date for Atlantean destruction from 9000 years to 900 years to fit in with the alleged approach of the comet Venus. There is no subterfuge intended, but Velikovsky's reasons for changing the date are weak, and an equally good case could be made for the original 9000 year figure, based on the time the Ice Ages ended.

Here is another seeming exaggeration (p. 86):

> The Choctaw Indians of Oklahoma relate: "The earth was plunged in darkness for a long time." Finally a bright light appeared in the north, "but it was mountain-high waves, rapidly coming nearer."

How could the writer, assuming he was the observer, have survived moun-
tain-high waves and lived to write about it? If he was not the observer, then
again we have hearsay evidence.

**Vagueness of Evidence.** Velikovsky believes that a side effect of the
near collision between earth and Venus was the change in seasons. Here is
some evidence (p. 133):

> "The breath of heaven is out of harmony. . . . The four seasons do not
> observe their proper times," we read in the *Texts of Taoism*.

This may simply refer to an extremely cold summer and a mild winter (as is not
unusual) and not to global catastrophe. Velikovsky does not admit of this
alternative possibility.

On the same point, Velikovsky draws this quote, in part, from IV Ezra 14:4
(p. 134):

> I sent him [Moses] and led my people out of Egypt. . . . I told him many
> wondrous things, showed him the secrets of the times, declared to him the
> end of the seasons.

The "end of the seasons" in this context could mean almost anything.
Nonetheless, it is accepted as evidence of Velikovsky's theories.

Consistent with the notion of an earth undergoing convulsions at the time
of the exodus, Velikovsky refers to Psalm 82:5 which states (p. 129):

> They walk on in darkness: all the foundations of the earth are out of course.

But could it not mean also that the Israelites were temporarily lost? And why
pick this particular psalm as evidence? Why overlook Psalm 104:5 which states
that the Lord is He

> Who laid the foundations of the earth that it should not be removed
> forever.

or Psalm 125:1:

> They that trust in the Lord shall be as Mount Zion, which cannot be
> removed, but abided forever.

These examples of ours suggest a stable and enduring earth, not a mobile,
cataclysmic earth. Here is still another biblical excerpt (Isaiah 40:4):

> Every valley shall be exalted, and every mountain and hill made low; and
> the crooked shall be made straight, and the rough places plain.

This is a near-perfect description of the geologist's uniformitarian concept
of gradation, or leveling, of the land surface by erosion. This is diametrically
opposed to the concepts of catastrophist Velikovsky. Now we do not believe for
one minute that the biblical scribe who wrote the above quote was thinking

about geology at all. The point is this: vague and ambiguous writings from other times and places can be selected and interpreted to prove almost anything you wish. This was the case with some of von Däniken's evidence, discussed in Chapter 7. This is not to say that historical evidence is worthless, but it must be used with care and impartiality.

We would agree with Dr. Velikovsky that his historical evidence shows violent episodes in the earth's past: earthquakes, fires, floods, landslides, volcanic eruption, unusual celestial phenomena, war, death and destruction. We have all of these things going on today around the world without relying on planetary collisions.

Ancient chronicles are full of these kinds of events because they were vivid, frightening, and had impact on the lives and property of people then and as they do now. Thus, these reports are not unlike what we see in our own daily newspaper. If we knew the world only through our newspapers, we would imagine the earth to be in constant turmoil. One would hardly expect ancient reporters to record for history that "everyone had a nice, quiet day and went to bed early."

Let us turn next to some of the scientific aspects of Velikovsky's theories.

## Geological Evidence

### General Background

Velikovsky uses a limited amount of geological evidence in *Worlds in Collision* because the book deals with only the most recent 3000–4000 years of earth history. Geology, of course, is concerned with all of the earth's 5-billion-year history. In Velikovsky's other book, *Earth in Upheaval*, he attempts to show the role of catastrophe throughout earth's prehuman history from evidence, as he expresses it, of "the stones and bones" found today as geologic formations. Thus, *Earth in Upheaval* is essentially a geologic exposition. However, we will restrict ourselves to a discussion of Velikovsky's geologic assertions as they are unfolded in *Worlds in Collision* because we believe the reader may be more familiar with *Worlds in Collision* rather than with *Earth in Upheaval*.

Chapter 2 of *Worlds in Collision* contains a general geological description of the earth and also a discussion of what are known as the "Ice Ages" but which geologists prefer to call the Pleistocene (the last 2 million years or so of earth history). Frequently throughout the book, mention is made of the hydrocarbons which rained down upon the earth from the comet's tail. According to Velikovsky, these hydrocarbons sank into the earth to form at least a part of the world's present supply of petroleum.

## The Planet Earth

Velikovsky's discussion here is hampered by the use of antiquated, out-of-date geological references whose average date of publication is 1914, and whose earliest is 1827. This was bad enough in 1950 when *Worlds in Collision* first appeared, but it is a less tolerable situation in 1977 reprints where the same references—unchanged, and averaging 63 years old—are used as supportive evidence. Old does not mean useless, but in a rapidly advancing science the old papers are often incorrect. The use of old references may explain the awkward handling of this excerpt from *Worlds in Collision* (p. 33):

> The propagation of seismic waves gives support to the assumption that the shell of the earth is over 2,000 miles thick; on the basis of the gravitational effect of mountain masses (the theory of isostasy) the shell is estimated to be only sixty miles thick.

The shell cannot be 2000 miles thick and 60 miles thick at the same time. We do not know what Velikovsky means by "shell." Perhaps he means crust, which is 20–40 miles thick. The crust certainly is not 2000 miles thick, and cannot be thought of as shell because this includes the mantle. This type of old, ambiguous information is not very helpful to the reader.

Also on page 33, Velikovsky informs the reader that "sedimentary rock is deposited by water." We do not mean to be picky, but sedimentary rock is also deposited by wind and ice. Further, this statement does not distinguish chemical precipitates, biochemical precipitates, and particulate settling from suspension.

Velikovsky sees difficult problems in geology in connection with lava flows (p. 37):

> . . . igneous rock, already hard, had to become fluid in order to penetrate sedimentary rock or cover it. It is not known what initiated this process, but it is asserted that it must have happened long before man appeared on the earth. So when skulls of early man are found in late deposits, or skulls of modern man are found together with bones of extinct animals in early deposits [we assume he means under igneous rock layers, as is clear from context], difficult problems are present.

We assume Velikovsky would opt for a catastrophist explanation to get out of this difficulty. It is not really necessary.

Igneous rock, as fluid lava, has been ejected from volcanoes repeatedly within historical times as well as long ago in prehuman history. People such as the Hawaiians and Sicilians are well aware of this. These flows often overrun villages and towns (Fig. 13-2). Therefore there is nothing mysterious about a hardened lava flow situated above human remains, either killed in the eruption or else long dead and buried beforehand. We do not understand the statement

Fig. 13-2. Lava flows overrunning human habitation. Igneous rocks overlying human remains are not unusual *(F. O. Jones, U.S. Geological Survey)*.

that these events "must have happened long before man appeared on the earth" because if man lies below lava flows then he already was on the earth.

Velikovsky also worries about coral reefs (p. 37):

> As far north as Spitzbergen, in the polar circle, at some time in the past, coral reefs were formed, which do not occur except in tropical regions.

This is not true. Coral reefs have been found living in the very regions he is talking about, and are known to be able to live in cold, relatively deep waters. We might add that, in fairness to Dr. Velikovsky, geologists were surprised at this too when it was discovered.


## Ice Ages

Velikovsky notes (p. 38) that "traces have been found of five or six . . . glacial periods." We would correct this to four glacial episodes. They are hardly "traces" because the ice scoured and profoundly modified the surface of the earth, streamlining hills and valleys in glaciated areas, and dumping millions of tons of glacial debris during their retreat. We know this to be true because glacial activity is still operative in many areas and we can make direct comparison of erosional and depositional processes. The effects of advance and retreat of continental glaciers in Europe and North America are well documented as having occurred during a nearly 2-million-year period. The coming of the ice was not an abrupt, catastrophic event. Flowing ice sheets two miles thick and hundreds of thousands of square miles in area are not made overnight.

As to the causes for the "Ice Ages," Velikovsky rejects the notions of varying heat from the sun or hotter–colder regions of space passed through by the solar system. We agree. Science also rejected these notions decades ago because they do not explain the rhythmic periodicity of ice advance and retreat. More sophisticated models are now being studied, but they are not predicated on catastrophism.

The question arises as to when the last ice age ended. Velikovsky employs 1911 geological estimates based on imprecise relative age dating which gives a minimum of 5000 years ago. Velikovsky arbitrarily suggests a minimum of 3000 years, fitting conveniently into the time frame of his cometary conflicts. As of the 1970s, based on radiocarbon dates, worldwide sea level changes, and other data, the date for the last glacial retreat was pretty well fixed in the magnitude of 12,000 years ago.

Finally, Velikovsky's weak grasp of geology is clearly shown by this statement (p. 40):

> Why did not the Ice Age touch this region [Siberia], whereas it visited the basin of the Mississippi and all Africa south of the equator? No satisfactory solution to this question has been proposed.

The Ice Age did *not* glaciate the Mississippi valley except in its most northern reaches. Nor was Africa south of the equator encased in ice during the past 2 million years, as is implied. Africa was glaciated no less than 200 million years ago and we fail to see what this has to do with collisions of planets. We might add that while Africa was being glaciated, the northern hemisphere was free of ice. Velikovsky confuses two hemispheric glaciations separated in time by many millions of years.

## Frozen Mammoths

Sensationalist authors have talked about these mammoths for years (Fig. 13-3). Velikovsky opts for sudden, catastrophic change, as do other writers such as von Däniken, because these mammoths are frozen in "blocks of ice" (p. 43) with plant material in their mouths. The mammoths are usually *not* frozen in "blocks of ice"—they are preserved in frozen *sediment,* a big difference.

There is nothing puzzling about such mammoths having undigested plant material in their digestive systems. These mammoths roamed along the fringes of the ice masses, and occasionally wandered up on the ice. Edible plants grew,

Fig. 13-3. A mastodon group *(Buffalo Museum of Science).*

as they grow today, along advancing ice fronts which intrude into temperate areas. It is not unlikely for a mammoth to have partaken of a vegetative meal, and within minutes or hours to have fallen into an ice crevasse. There is no evidence of a mass calamity enveloping all mammoths. The report of mammoths standing in a frozen position is drawn from the singular case of a mammoth that fell into a narrow ice crevasse and died in an upright position. Usually, no mention is made of the fact that this mammoth is partially reclined, and has a broken leg and tusk. The tale of a mammoth, calmly standing and minding his own business, munching buttercups, when overcome in his tracks by some kind of fast freeze, is pure fiction.

## The Hydrocarbon Deluge

If there is any bastion of the geologic domain Velikovsky cares to invade and be greeted by enthusiastic opposition, this is it. Velikovsky claims, and it is a crucial part of his whole theory, that the tail of the comet Venus was loaded with hydrocarbons (oil) which, during the encounter with earth, rained down upon the planet to produce fiery deluges. In the process, much of this oil percolated into cracks and fissures, and sank into the earth to become part of the world's oil supply. Velikovsky notes that countries such as Egypt, Iraq, and Saudi Arabia, where legends of "rains of fire" persist, are also oil-producing countries.

We are not concerned here with whether or not the tail of the Venusian comet *could* contain hydrocarbons (although in the next section we will be). For the sake of argument, let us say it *did* contain hydrocarbons. As Velikovsky expresses it (p. 73):

> The rain of fire-water contributed to the earth's supply of petroleum; rock oil in the ground appears to be, partly at least, "star oil" brought down at *the close of world ages*. . . .

All the knowledge and experience of thousands of oil-finding geologists for more than a century militate against this view. We do not mean to advance an argument here based on "authority," but geologists have been highly successful in predicting the occurrence and entrapment of oil based on certain principles that have nothing to do with a rain shower of oil onto the planet. These principles are based on other factors we would like to summarize briefly.

Oil is generated by complex processes from organic-rich sediments such as black shale. The process continues today. Upon attaining a fluid condition, the oil migrates to porous-permeable rocks, chiefly (98%) sandstone, limestone, or dolostone. These rocks must be so structured as to form a "trap" (Fig. 13-4). All traps possess an *overlying* cap rock completely impermeable so as

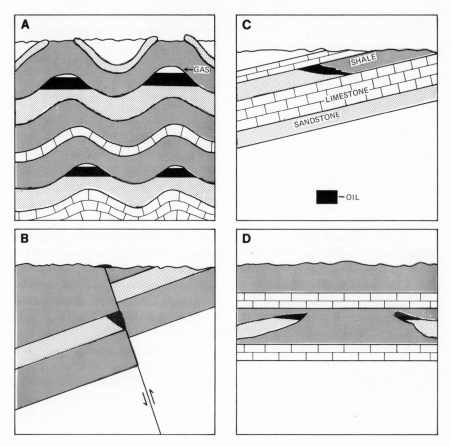

Fig. 13-4. Typical oil traps. The occurrence of oil and gas is not consistent with Velikovsky's theories.

to confine the oil because oil migrates *upward,* not *downward* (more precisely, it migrates to a position of least potential energy). The bulk of all oil is found associated with ancient sediments whose depositional environment was near-shore, shallow, and marine. Thousands of studies by thousands of geologists make these ideas abundantly clear. A geologist who wanted to keep drilling in nonmarine sediments would be fired by his company.

Given this information about oil, we would like to make the following points concerning Velikovsky's hydrocarbon thesis (we do note that Velikovsky does not claim *all* oil came from the comet's tail, only part of it; we would insist that *no* oil came from the tail of Venus):

1.  If oil rained down upon the earth 3500 years ago in the impressive

quantities described by Velikovsky, why are not all or most surficial rocks around the world impregnated with the stuff? There should be at least traces detectable by chemical analysis. There are not.

2. If oil percolated down from the surface, why should it seek out only rock units of shallow marine origin in which to reside? Why not all kinds of rocks? If this were the case, we would find oil by guesswork and prayer. Oil is not found in this way.

3. The oil would have had to migrate downward, then turn around and migrate upward in order to reach the traps where oil is found. This violates simple physical laws.

4. If these hydrocarbons were burning as they fell, little would remain to migrate below, except conceivably a tarry residue, which could not migrate.

5. Oil and natural gas are closely associated. When oil is brought to the surface, much gas "fizzes" out (like soda pop) because of the change to the lower pressure at the surface. If we are to believe Velikovsky, oil rained from the skies, descended to high pressures at depth, miraculously picked up natural gas (also a hydrocarbon) somewhere along the way, and dissolved it into the oil, to be later released as the oil was produced. Again, this defies physical laws.

6. If Velikovsky were correct, we should find oil in igneous and metamorphic rock such as comprise the vast Canadian Shield area and other shield areas around the world. Again, we do not. They are restricted to sedimentary rocks. When oil is found—rarely—in igneous or metamorphic rocks, it is a subject of special interest to the geologist, and in all cases can be traced to nearby sedimentary rocks.

On the basis of the preceding facts, the assumption that hydrocarbons rained down from the heavens 3500 years ago and formed part of our oil deposits is without logical or scientific foundation. We believe that any one of the above six points, by itself, is sufficient to discard Velikovsky's notion. Many, if not all, of Velikovsky's historical references to a "rain of fire," were probably due to the actual fact of fiery cinders or particulate hot lava violently expelled during volcanic eruption and "raining" down upon the populace. Such would be memorable and worth writing about.

## The Darkness

Velikovsky presents repeated references to darkness enveloping the earth for long periods of time. He points out in *Worlds in Collision* (p. 139) how the eruption of Krakatoa in 1883 darkened the skies even at noon. This eruption, as Velikovsky correctly points out, caused explosive sounds heard 3000 miles away in Japan (and Australia). We might add that the unreal darkness was also observed during the eruption of Vesuvius which destroyed Pompeii in A.D. 79. But we think that Velikovsky's main point is that if a single eruption could blot out the sun and cause darkness at noon, imagine how really dark it could be, and for how long, if the earth were involved in near collision with Venus, with clouds of opaque smoke and dust emitted from many, many eruptions infesting the atmosphere of the earth plus similar contributions from Venus.

This translates, in Velikovsky's view, into a 40-year period of darkness (or at least a twilight situation) coinciding with the same period during which the Israelites wandered in the desert. Maybe we are wrong, but if the earth were enveloped in darkness for 40 years, we think this would spell the end of humanity, and most if not all of life on earth. The cataclysm, as we understand it, had already wiped out most of humanity. The deficiency of light would preclude the growing and harvesting of crops anywhere in the world. Herbivores would starve to death; carnivores would eat carnivores to the end. It would not take long. Whatever food supplies people had hoarded would soon be exhausted. Human life would perish.

But Velikovsky has a way out of this dilemma. Manna. The same rain of hydrocarbons from the comet's tail also produced edible carbohydrates which kept the Israelites going for 40 years while in darkness. The "manna from heaven" may not be miraculous. Manna, several varieties of it, is still produced today by certain trees and bushes in the Middle East. It may be released on the wind and fall to the ground. It is edible and even considered a delicacy in some quarters, according to a lengthy article on the subject in the *Encyclopaedia Britannica*. An abundant manna fall took place in Turkey in 1891.

If manna sustained the Israelites, how were the blacks in Africa, the Chinese, and the Indians in America keeping alive during the 40-year darkness? We do not hear of manna falls in these areas. If not, they would certainly have perished; in which case the entire human race ought to be descendants of the Israelites. But it is not. The manna explanation does not sustain Velikovsky's argument. Whatever darkness there was at the time of the exodus, it seems likely that it was not worldwide, or sufficiently intense to prevent plant growth.

We would not disagree with Velikovsky's historical evidence that something tragic and profound *did* happen at that time, causing earthquakes, floods, darkness, death, and destruction that gave rise to the many writings that describe it. It was a singular and spectacular event that profoundly affected the Mediterranean area, the source of the most prolific and articulate descriptions cited by Velikovsky. We suspect that this was the mighty eruption of Thera (see Chapter 10), greater than Krakatoa. Little of this catastrophe was known at the time Velikovsky was gathering his evidence or he may well have modified his theories.

Here, we would like to say a few words about Thera.

## Thera

To appreciate Thera's capabilities as a volcano, let us first consider what happened when Krakatoa, smaller than Thera, erupted in 1883. Here is a description from a recent geology textbook (*Physical Geology* by Cazeau *et al.*, p. 96):

> On August 26, loud explosions were heard 100 miles away, and dense clouds of ash and pumice shot 17 miles into the air. Along the coast of Java and Sumatra, darkness fell as volcanic clouds shut out the sun. This darkness lasted two and a half days. Torrential ash-laden rains added to the turmoil. On August 27, Krakatoa reached its peak of explosivity. The sounds of a series of detonations were heard in Australia, 3000 miles away. At the same time, volcanic debris was thrown [50 miles] into the sky. The finer particles, riding on stratospheric winds, encircled the earth and took two years to settle. It is estimated that 4–5 cubic miles of rock debris were blown into the air during the paraxyms that pulverized Krakatoa, with an energy release matching that of the most powerful hydrogen bomb . . . the reverberations of Krakatoa unleashed a *tsunami,* a large sea wave that attained a height of 120 feet from base to crest as it crashed against the coasts of Java and Sumatra, and swept 36,000 people to their deaths.

Here we see the main ingredients Velikovsky calls for in his Venus–Earth encounter: darkness, deafening explosions, unusual rain, shaking of the earth, gigantic waves and flooding.

Now, take a look at Thera (what's left of it), 60 miles north of Crete in the Aegean Sea. It is a crescent-shaped volcanic ediface rimmed by steep crater walls 1000 feet high. These walls extend down into the water for at least an additional 1100 feet. The ancient crater has a perimeter of 18 miles, enclosing on three sides a very large bay.

This was the locus of a catastrophic eruption in ancient times exceeding that of Krakatoa. The time of the eruption has been pinned down by means of radiocarbon dating and archaeological evidence. The eruption took place during the 15th century B.C., the same time that Velikovsky believes a comet grazed the earth.

We can expect the same effects, but on a somewhat magnified scale, resulting from Thera's "big bang" as we described for Krakatoa. The effects must have been awesome: the entire Mediterranean region shaken and cast into darkness; explosions like thunder heard for thousands of miles; violent rainstorms, the water impregnated with red volcanic ash; red-hot cinders scattered on the wind; giant tsunamis surging back and forth in the darkness, withdrawing water from the shores and then returning to flood miles inland. There must have been complete terror. It is now known that the end of the Minoan civilization on Crete came as a result of this eruption.

No wonder Velikovsky found so many references to these convulsions of nature in ancient literature. We think Occam's Razor applies here in the case of the geologic evidence. A much more plausible explanation exists for cataclysmic events described by ancient writers than the awkward theory of Velikovsky. Hypothetical comets may have come and gone, but Thera is there for all to see.

We would like next to consider some of the data from astronomy, astrophysics, and the space programs.

## Cosmological Considerations

### The Solar System

It is difficult for us to reconcile Velikovsky's assertions about the solar system with what is known about it today. Astronomers have long maintained that any rational account for the origin of the solar system must recognize and reconcile the observations that all the planets are moving in the same direction around the sun, in nearly circular orbits and in almost the same plane. There is an intrinsic order here. It would be, in our view, totally impossible that Venus and Mars both could, willy-nilly, hop about the solar system 3500 years ago, nearly colliding with one another and with other planets, and finally come to positions fitting in so harmoniously with the rest of the solar system. Velikovsky's theory demands a great deal of—what—faith?

### Comets

Even in our earliest readings of *Worlds in Collision* many years ago, Velikovsky's use of the term "comet" troubled us. Venus is presently a planet with a density (4.9) almost as great as that of the earth (5.5). We thought comets were bodies lacking in substance, so fragile that the sun's radiation will "blow" a comet's tail around so as to precede the comet's head in the direction it is moving (Fig. 13-5). We tend to agree with L. Sprague de Camp, who says with reference to Velikovsky (*Lost Continents,* p. 94):

> . . . the theory is ridiculous from the point of view of physics and mechanics. Comets are not planets and do not evolve into planets; instead they are loose aggregations of meteors with total masses less than a millionth that of the earth. Such a mass—about that of an ordinary mountain—could perhaps devastate several counties or a small state if it struck, but could not appreciably affect the earth's orbit, rotation, inclination, or other components of movement. And the gas of which a comet's tail is composed is so attenuated that if the tail of a good-sized comet were compressed to the density of iron, I could put the whole thing in my suitcase!

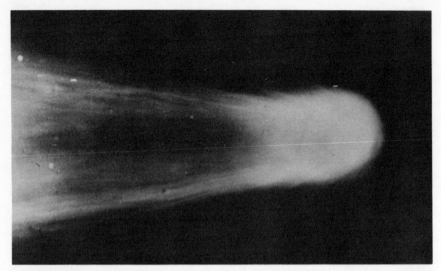

Fig. 13-5. Halley's Comet. Comets are diffuse and have little mass. Venus could not be construed as a comet *(Hale Observatories).*

Maybe we are just quibbling over terminology and we should simply call Venus an astral body. It appears that Velikovsky called it a comet because early writings refer to Venus as having "hair" or a "beard." Whatever, a related problem is the density.

Venus is said to have been spawned out of Jupiter. Velikovsky's evidence is exemplified by the Greek legend (p. 177) that "She [Venus] sprang from the head of Zeus–Jupiter." It is our understanding that Venus came from the foam of the sea and it was rather Athene who sprang from Jupiter. Be that as it may, how this could have been known in ancient times mystifies us. Without telescopes, Jupiter is a point of light like so many others in the night sky, if and when it is visible. We do not understand how ancient astronomers could have seen a splitting off of a relatively small body from the giant Jupiter.

In any case, let us assume it happened. Jupiter has a density of 1.3. Present-day Venus has a density, as we have noted, of 4.9 and a diameter of 7700 miles. Given the present mass of Venus, we are required to eject a large ball of material from Jupiter with a density of 1.3 that would have to be at least four times the size of the earth. This mighty ball must then shrink and condense down to its present dimensions and form a solid crust and a density of 4.9, and we have only 3500 years in which to do it! To compare, there is a large body of igneous rock that was intruded as a molten mass beneath the state of Idaho about 70 million years ago. This mass is infinitely small by Venus standards, yet, even after 70 million years, it is still cooling off!

To repeat, Velikovsky's theory demands a great deal—the abridgment or modification of some very elementary rules of science.

## Planetary Composition

The presently known compositions of Jupiter and Venus do not fit into Velikovsky's scheme at all. At the time Velikovsky developed his cosmological ideas. Jupiter was thought to consist of ammonia ($NH_4$) and methane ($CH_3$). Methane is a hydrocarbon, and a case could be made, albeit tenuous, for a rain of hydrocarbons upon the earth if Jupiter's offspring passed close to the earth. Deep-space probes of Jupiter by the United States indicate rather strongly that Jupiter is *not* made up of methane and ammonia as previously thought, but is instead a swirling ball of hydrogen. It is impossible to produce a comet Venus rich in hydrocarbons from nothing but hydrogen.

Related to the problem is the present composition of Venus itself. Space vehicles have passed through the Venusian atmosphere and landed on its surface by parachute, carrying a variety of sophisticated instruments. Bearing upon the cosmological ideas of Velikovsky is the discovery that the atmosphere of Venus is dominantly carbon dioxide and its crust, although we are uncertain of rock type, is granitic in composition. Granite is essentially quartz ($SiO_2$) and feldspar ($KAlSi_3O_8$).

Velikovsky's cosmology must explain how a ball of hydrogen (Venus) can acquire carbon on its way to earth, cast hydrocarbons thereon, continue through space and somehow pick up a load of oxygen to make carbon dioxide, get rid of its excess hydrogen, and then fall into its present harmonious orbit. Also, it would have had to acquire the necessary ingredients to make sulphuric acid ($H_2SO_4$) and hydrofluoric acid (HF), present in the Venusian atmosphere as well as the necessary $SiO_2$ and $KAlSi_3O_8$ or related materials to construct the surface that space vehicles landed on. It is a tall order.

Adherents of Velikovsky's theories have an answer to this. They say that during close encounters with Earth, Mars, etc., Venus picked up what it needed. For example, Venus picked up the oxygen from our atmosphere necessary to make carbon dioxide. This sounds good, but our atmosphere is three-fourths nitrogen. Are we to believe that in its passage close to us, Venus had the uncanny ability to sort out the unwanted nitrogen and take only the oxygen? We think even the most enthusiastic adherent of the Velikovsky cosmology would tremble at the thought.

## Final Thoughts

During the 1950s, Immanuel Velikovsky offered the world an intriguing and imaginative theory. We think all theories should be welcome, right or wrong, because they stimulate thought, debate, and investigation, and thus science advances. There are many examples in the history of science to show this. Unfortunately, Velikovsky's theories became an emotional issue. Two camps quickly emerged: one, a band of sneering skeptics using the tactic of ridicule; the other, a band of fervent adherents searching for rebuttals to each new scientific finding, and donning the martyr's cloak of persecution. Neither camp serves science and truth.

We have tried, in our own way, to look at Velikovsky's theories and evidence fairly and without prejudice. It should be obvious that we do not believe Velikovsky's theories are tenable at this time. This is by no means a condemnation of Dr. Velikovsky. We respect him for his tireless energy, dedication, and wide-ranging mentality. We need people like Velikovsky, whether they are right or wrong. They stir the fires of science and keep them kindled.

We are reminded of that great geologist, A. G. Werner, who more than 150 years ago announced a revolutionary theory that all of the earth's rocks had been precipitated from a universal ocean girdling the globe. He and his followers were called Neptunists. Werner was eventually proved wrong, but not until numerous geologists had deserted their armchairs to go into the field to try to find evidence that he was either right or wrong. Geology did not retreat; it advanced. If Werner had a fault, it was in becoming so enamored of his own theory that he ignored other viable facts and more plausible alternatives to what he had entrenched in his mind.

# CHAPTER 14

# Noah's Ark

*Today more than one and a half billion people—Christians,*
*Jews and Moslems—know the story of Noah's Ark.*
—BALSIGER AND SELLIER

This story, as well as other flood stories among other cultures, is believed by many people. Those who adhere to a literal translation of the Bible insist that the entire earth was covered by water. To what extent is this concept supported by science? What is the evidence, if we are to approach this concept scientifically? The notion of an entire planet rapidly submerged by a universal ocean is certainly catastrophist. Was the earth so visited by such a catastrophe? There are those who claim that it was, and invoke scientific evidence, they say, to support it. Furthermore, it is claimed that—as ultimate and undeniable proof—the remnants of the wooden ark still lie on the slopes of Mount Ararat (in Armenia) and its wood has been obtained, studied, and found to be authentic.

The purpose of this chapter is to examine the entire question of a universal flood, the evidence that is presented to support the hypothesis, and the probability that the literal biblical tale is true. We approach this topic with the attitude that (1) it is a catastrophist idea, and (2) the Bible has a root of historicity nonetheless. We do not wish to offend anyone's personal religious beliefs, but yet the findings of science—especially archaeology and geology—have meritorious data that should be taken into account in assessing the meaning of the biblical story of a universal flood.

## Evidence for a Universal Flood

The universal flood of Noah was supposed to have taken place about 2448 B.C., according to Hebrew chronology. Another date suggested is 2345

B.C. The Greeks say it took place about 3050 B.C. The flood covered the entire earth, and according to Genesis 7:20 the highest mountain (presumably Mount Everest) was covered by water with 22½ feet to spare.

An initial problem is the source of such tremendous amounts of water to extend the earth's radius by 5.5 miles (height of Mount Everest = 29,028 feet + 22½ feet). By our calculations, more than a billion cubic miles of water must be added to the oceans to cause sea level to rise 5.5 miles. In contrast, note that if all the ice sheets of Greenland and Antarctica and elsewhere were to melt, sea level would rise less than 200 feet.

There are insufficient quantities of water in the atmosphere to generate a billion cubic miles of rain. Even if there were such amounts of water, it also must be explained what happened to this water once the flood receded. Some flood believers get around this by appealing to Genesis 7:11 which states that ". . . the same day were all the fountains of the great deep broken up, and the windows of heaven were opened." This is interpreted to mean that a major source of water was the subsurface of the earth. Vast quantities of underground water are visualized as spewing out of the earth to add to the rainwaters, and once the flood was past, returning the same way it came. Although there are indeed large amounts of subsurface waters, we know of no mechanism that would expel it all to the surface for a limited time (less than a year) and then allow it to drain back into the subsurface. Groundwater simply does not behave that way. It would require some innovation of the miraculous.

But let us say it happened. Then what would be the condition of the earth's surface today, after being ravished by miles-high waters only 5000 years ago? Even the highest mountains would show the erosive effects of moving currents. They actually do not. Regions glaciated 12,000 or more years ago (during the Ice Ages) would bear the imprint of modification by moving waters. Again, they do not. The open scars of glaciation still remain in pristine condition, whatever the elevation.

Two authors who claim that the flood of Noah is supported by scientific evidence are Dave Balsiger and Charles Sellier, Jr. The major theme of their book *In Search of Noah's Ark* (1976) is a marshaling of the data of science in favor of their hypothesis. They believe that the great flood would have left extensive sedimentary deposits around the earth. So far, so good; but then they say (p. 40):

> Such a flood would have to deposit tremendous amounts of sedimentation [*sic*] throughout the world. And it has been scientifically estimated that more than 75 percent of the earth's surface is sedimentary in nature. . . .
> In the northern Rockies, scientists have found well preserved fossilized trilobites and other delicate insect fossils. . . . They did not expire slowly, but abruptly, from an unexpected catastrophe, such as a great flood.

It is correct that 75%, perhaps more, of the earth's surface features sedimentary rocks. But this figure includes all sedimentary rocks deposited over a period of more than 600 million years, and under a variety of environmental conditions such as lagoonal, open ocean, river, swamp, desert, seas, and so on. The implication that they were deposited more or less simultaneously in catastrophic fashion is an absurdity. Note the reference to trilobites, fossils found in these rocks. Trilobites have been extinct for more than 200 million years. Also, they were not insects. Their conditions of preservation do not in any way suggest catastrophe. The rocks authors Balsiger and Sellier refer to also include the entombed remains of dinosaurs. They have been extinct for more than 70 million years. If these sediments, separated by long periods of time, had been deposited during the biblical flood, then Balsiger and Sellier would have to accept the thesis that dinosaurs existed at the time of Noah. In which case they should have been in the ark along with other reptiles.

A universal flood would have left extensive unconsolidated sediments as a veneer over most of the land areas of the earth. There is no evidence whatsoever of this veneer (Fig. 14-1). For example, surficial rock over much of

Fig. 14-1. These sedimentary rocks offer no proof of Noah's flood; they are millions of years old *(Atlas and Glossary of Primary Sedimentary Structures).*

Canada is of Precambrian age—greater than 600 million years old, and with little sediment on top of it.

## Occupants of the Ark

### Space Problems

The ark would have had to hold many animals of diverse types. According to Bible scholars, two of every kind of animal came to the ark instinctively, so that Noah was not required to track them down actively. We must admit to a certain wonder that only a select pair of each animal species out of several thousand individuals responded to this instinct for survival and joined the worldwide procession to the portals of the ark.

The Bible states exactly the size of the ark in cubits. Translated to feet, the ark was $450 \times 45 \times 75$ feet, with a volume of 1,518,750 cubic feet. The next question is, how many creatures must fit into this available space? Most analysts include only birds, reptiles, and mammals. Fish, amphibians, insects, and plants are usually neglected, although the Bible states specifically that (Genesis 6:17) ". . . everything that is in the earth shall die."

If we consider only birds, reptiles, and mammals, and assume that no new species have been created during the past 5000 years, then we have the following conservative figures: 3500 species of mammals plus 6000 species of reptiles plus 12,000 species of birds equals 21,500 total living species; doubling that gives 43,000 individuals on the ark.

If we divide the figure 43,000 into total cubic feet available in the ark, we arrive at about 35 cubic feet per animal. This may sound like ample space, but it only approximates the space of a human coffin ($6 \times 2 \times 3$ feet). We can assume that pairs of the same species would be domiciled together aboard the ark, so that each pair has 70 cubic feet of space, equivalent to a square box four feet on a side. Naturally, large animals such as elephants, camels, giraffes, and rhinos would have larger quarters, and smaller creatures such as birds would have to content themselves with much more cramped quarters.

Our assumptions here, unfortunately, have failed to take into account several other space factors, which suggest less room for the ark animals:

1. Considerable space would be taken up by the cages, partitions, and passageways of access to the animals.
2. Storage of food and water necessary for 43,000 creatures would require significant amounts of space, considering that they would be aboard the ark for at least 275 days. There would also need to be some interim storage for animal waste products.

3. Some space would be needed by Noah, his family, their furniture and other possessions, food, and water.
4. Some of these animals doubtless would reproduce during the long voyage, taking up still more space, food, and water.

Taking into account all of the above factors, we would guess that the ark could accommodate all of its creatures, but living conditions would be incredibly crowded, unsanitary, and oppressive, perhaps not unlike concentration camps of World War II.


## Food and Water

We have alluded already to the amount of space which would be taken up by the food supply. There are a few other practical considerations involved here. The diversity of diet is one. Would not meat be necessary to feed the carnivores? How would fresh meat be kept for nearly a year? We would guess some animals were kept aboard to be slaughtered periodically. Exotic and specialized foods would be needed by such varied creatures as the Queen Victoria crowned pigeon, the bald eagle, the scarlet ibis, moles, and spiny anteaters, to name a few.

If each creature, on average, consumed half a quart of water each day, then the minimum amount of water needed to be stored in the ark for the duration of the voyage would be 1,263,125 gallons. This is a formidable quantity to gather and store in the ark. Would Noah have been aware of how much water he would need? Ships today take aboard supplies sufficient for voyages of known duration. Noah had no guidelines in this respect, insofar as we know.

These great difficulties have been easily resolved by flood people. They suggest that most of the animals on board the ark went into hibernation. Thus, food and drink requirements would be quite minimal. There is no evidence in the Bible that we can see that this happened. It is a speculation, and in our opinion, a weak one. Animals removed abruptly from their native habitat, placed aboard a moving, unstable ship under crowded and unfamiliar conditions are not likely to lay down and take protracted naps.


## Caring for the Animals

There were eight human beings aboard the ark. The fate of 43,000 creatures was in their hands. These animals needed to be fed, watered, and cleaned. It works out to each of Noah's family (including Noah himself) being responsible for 5375 animals. Thus each member of Noah's family, if we assume generously, would need only four hours sleep each night and could spend the remaining 20 hours of each day caring for the animals at a rate of 270 animals each hour. Noah's family would have needed to step lively.

We might interject that the care of animals, as zookeepers, is a skilled occupation (Fig. 14-2). It requires knowledge, stamina, patience, insight into animal behavior and needs, and a natural rapport with a variety of creatures from anacondas to aardvarks—a knowledge of their habits, diet, special needs, dangers to health, and so on. Could Noah's family have had these skills? Many, perhaps most, of these creatures they would be meeting for the first time aboard the ark. It would be, again, miraculous that Noah's family would possess these abilities and the wherewithall to care for the animals at a rate of 270 animals per hour. It seems impossible.

## Further Thoughts

A crowded ark, limited food and water, animals without exercise, an incapability to care adequately for a large number of unfamiliar creatures lead inevitably to the conclusion that we are talking about an all-but-impossible

Fig. 14-2. Zookeeper at work. The care of animals is a skilled profession. Did Noah and his family have these skills? *(San Diego Zoological Society).*

situation. Under the conditions we describe, it would be most surprising if even half the animals survived the 275 days aboard ship. In which case, we would not see the diversity of life that actually exists here on earth. We conclude that the saga of Noah's ark, with respect to the particulars depicted in the Bible and by biblical flood believers, does not adhere to reality and could not have taken place as described, according to science and common sense. If one wishes to invoke Divine influence, then of course all things are possible and our own analysis becomes nugatory.

## Descendants of the Ark

One of the greatest pieces of intrinsic evidence against the literal story of Noah's ark is the present type and distribution of races on the earth. If the biblical story is to be believed, then all humans on earth today are descendants of Noah's family. This includes Chinese, Japanese, pygmies in Africa and other blacks, blond Scandinavians, eskimos, copper-skinned Indians, Hebrews, Arabs, various Caucasians, Polynesians, and Australian aborigines. The list goes on and on.

There is no problem in repopulating the globe beginning 5000 years ago. With the exponential increase in population, Noah's family would be capable of supplying the present planetary population and then some. However, there is the sticky problem of race differentiation. This takes time—far more than a few thousand years. It is difficult for us to believe that this could happen. Furthermore, many cultures and races can trace their origins back almost to the time of the flood. The Egyptians are an example. We would have to presuppose that one family with offspring from the same parents, and of common genetic characteristics, managed in a few thousand years to spawn black, white, red, and yellow races with associated subvarieties. Such a result can only be construed as a fairytale.

## Ararat and the Ark

According to the biblical account, Noah's ark came to rest on or near the summit of Mount Ararat as the waters receded. Mount Ararat is situated in Turkey near the Russian border, and rises to 16,984 feet above sea level. Snow and ice occupy the summit area. It has been climbed on numerous occasions, and is not considered a difficult climb by professional mountain climbers. There are those who claim that the ark is still there, encased in snow and ice near the summit. According to some, the ark has been seen, even entered during periods of ice recession, and wood from the ark has been brought back. This wood has been scientifically dated, the dates fitting into the chronology of biblical history.

Historical sightings of the ark are inconclusive, from the scientific point of view. The source of many eyewitness accounts are early churchmen with a special ax to grind, such as John Chrysostom, who said in a sermon (quoted by Balsiger and Sellier, p. 76):

> Do not the mountains of Armenia testify to it, where the ark rested? And are not the remains of the ark preserved there to this very day for our admonition?

The questions are merely posed. No evidence is offered. Did John Chrysostom himself see the ark? He does not admit to this seemingly important question. Nor does he say who did. This is, at best, hearsay evidence.

Among the array of sightings we find it asserted by Balsiger and Sellier at various points in their book that the ark (1) is encased under 30 feet of ice, (2) is resting in a lake, (3) is perched on a narrow ledge along a precipitous cliff, (4) is lying in the open in a glaciated valley, and (5) has been broken up by ice movement and its remains scattered over a wide area. It is clear that these sightings are contradictory, and cannot be classified as evidence.

Consider the situation from a geological viewpoint. The summit area of Mount Ararat is one of snow accumulation and glacier formation. Small glacial tongues move downslope to lower elevations, picking up boulders, sediment, or anything else in their paths. Moving glaciers have been compared by geologists to bulldozers.

We believe that if the ark had come to rest near the summit of Ararat 5000 years ago, it likely would have shifted by glacial movement to lower elevations long ago. To at least some extent, the ark would have broken up, the wood strewn about on the lower slopes of the mountain, easily accessible even to those who are not mountain climbers. No such quantities of wood, let alone the remnants of the ark itself, have ever turned up insofar as we are aware.

What we fail to understand is why a single large boat presumably perched on one isolated mountain peak should elude its pursuers for 5000 years. If it is visible, aerial photographs from the many aircraft that have overflown the summit should be in plentiful supply. They are not. Photographs offered in evidence are dark, obscure, and require an active imagination to distinguish the rectangular ark from the many similarly shaped natural features that adorn the summit and upper slopes. Why should not a determined effort settle the question once and for all by landing search parties by helicopter on the upper slopes? Such an effort has been suggested, but questions about the "political situation" seem to rule this out, because of the proximity of the mountain to the Russian border. Is it really that formidable a problem to work out? We don't know.

The existence of the remains of Noah's ark on Mount Ararat can be considered only a hypothesis. In the final analysis, this hypothesis has yet to be adequately tested.

The ark, according to the biblical story, was constructed out of "gopher-wood." No one knows what gopherwood is, but it is thought to be either cedar or white oak. Wood allegedly from the ark has been subjected to testing to determine its age, according to Balsiger and Sellier. These authors devote considerable discussion to age dating of wood, presumably from the ark, discovered by Fernand Navarra in 1955. If the wood turned out to be approximately 5000 years old, it would be solid evidence in favor of the ark theory.

It is regrettable that the Navarra wood was subjected to outmoded and imprecise methods of testing such as (1) degree of lignitization, (2) degree of fossilization, and (3) gain in wood density. Estimates of ages from these kinds of tests can only boil down to a matter of opinion. This is why they were abandoned by scientists long ago in cases where the age of a sample was of critical importance. Be that as it may, those who analyzed the wood by these methods reported that the age of the wood "varies around 5000 years," "a period dating to a remote antiquity," and "estimated to be 4000 to 6000 years old." These estimates seemed to fit in with the presumed age of the ark.

The wood was then subjected (courageously) to radiocarbon dating, the preferred modern method of archaeology and geology. Several dates from three different laboratories gave an age range of between 1250 and 1700 years—far too recent to fit into the ark theory (Balsiger and Sellier, pp. 185–186). Undeterred, these authors then devote several pages to show how unreliable the carbon-14 method is, and construct a speculative, awkward theory of atmospheric evolution to explain away the carbon-14 results and at the same time show how it was that Noah and other biblical patriarchs could live several hundred years. We wonder if the carbon-14 method would have been so meticulously scrutinized as to its reliability were the dates in the range of 5000 years, and thus fitting into ark chronology. Without malice, we think the dates would have been accepted without hesitation.

## Real Floods

Our foregoing analysis of the story of Noah's ark and a universal flood casts serious doubt on the actuality of such a worldwide flood. The booking of so many varied and abundant animal passengers on this ark also seems difficult to accept. But does this mean that something comparable to the biblical account could *not* have happened? Not necessarily.

Even in modern times, violent and destructive floods by the Tigris and Euphrates rivers have occurred. This is the region of Noah. Such floods also occurred in ancient times. Flood-deposited strata of silt, sand, and clay are known from this area, and date back to 3000–4000 B.C.

One flood, greater than all others, can stick in the memory, and become known as "the" flood to the survivors. In the same way, to use a modern example, the great blizzard that struck Buffalo, New York, during the winter of 1977 quickly became known as "the" blizzard, so vivid was this snowfall in an area whose inhabitants shrug off "ordinary" blizzards of two feet or so.

Could we not then reconstruct a scenario of a great flood 5000 years ago in the area of Mesopotamia? One family, that of Noah, more alert or cognizant of the warnings of nature, anticipate a great flooding of the local rivers. They build a boat, or perhaps enlarge and refurbish an existing boat, and stow away supplies. They allow space for their livestock, a natural impulse.

The flood strikes, a disastrous and memorable one, with extensive loss of life in the region where they live, which is regarded as their "world." After several days, the ship grounds upon some high ground or hill. With time, the legend of this event is handed down as the familiar Noah's ark story. It would be easy to imagine the flood as the act of an angry God, sparing those who were deserving.

Perhaps hard answers someday will be forthcoming that explain the true nature of the biblical story. Until that time comes, the story of Noah's ark will remain another of the earth's great mysteries.

# Epilogue

In 1930, famed globe trotter Lowell Thomas concluded a chapter of *India: Land of the Black Pagoda* with this thought (p. 190):

> That it is not in accord with facts is no reason in the public mind why it shouldn't be true.

Our reading of mysteries that captivate the popular imagination convinces us that Thomas's observation is a major theme in human curiosity and applies equally well today. In the preceding chapters we have seen how, in disputed areas, there is a deep division between what might be called the scientific establishment and the nonprofessional. There is a tendency for each side, in its own self-interest, to look into the other's mind to explain the dichotomy. Amateurs sometimes see professionals as striving at all costs to preserve scientific dogma—a mental hardening of the arteries. Professionals, on the other hand, often see their adversaries as the unorthodox who ignore empirical methods of science and embellish a few selected facts with fantasy. Even without such prejudicial feelings, it is a complicated matter. There are many shades of outlook on both sides. Amateurs are not necessarily unscientific, and on the other hand the Ph.D. does not confer either total rational thought or a rigidly arrogant stance.

Other considerations we have touched on are truth in journalism and respect for honest opinion. Often the press, we have noted, rushes to publish as fact that which should be considered as theory still in need of supporting evidence. Honest opinion should always be open to consideration by amateur and professional, but if published prematurely as fact, such opinion is open to attack from all sides.

We have taken a spotty tour of the world and a few of its mysteries. Our criterion of selection has been to look at mysteries that have lured large numbers of people, but may have received little response from the scientific community. In some cases, the mysteries may be anecdotes or tales probably born of a fear of the unknown. In other cases, we suspect that man has had a strong hand in creating fictional mystery and providing his own answers, occasionally by deception. Many great mysteries of the archaeological past are the result of irresponsible speculations based on irrelevancies and facts removed from context. Other mysteries ignore even the most elementary data of geology and the other physical sciences.

We have tried to show that often what seems to be a mystery is not mysterious at all. An example of this is illustrated by the sculpture in Figure E-1. It bears the stamp of imposing symbolism. It stands today among other large statues on a hillside above Ashford Hollow, New York. If it were found in the distant future as a ruined monument, we might easily draw some conclusions largely based on inference from archaeological detail. Its appearance and general setting might indicate an abstract and fully developed alien style—a prima facie case for a god in the guise of a symbolic figure. We could easily believe that the image represented an entire faith and consciousness of a mysterious, departed race. One could imagine groups gathering periodically to fulfill their spiritual aspirations at the foot of this awe-inspiring idol. In actual fact, this piece of sculptural art is not a religious shrine, unless the reader considers the inner feelings experienced by the creative sculptor to be "religious." The product itself is an appurtenance of an educational and recreational program—an attempt by one man and his friends to involve a 20th-century community in the visual and performing arts. Viewed as a solitary figure, it is the epitome of the mysterious. In its context, with the motivation of its creator known, it is no mystery at all.

We pose the question to show that, too often, we are prone to interpret sculptures, pyramids, and stone circles based on their appearance today as though their significance is mysteriously locked in their present physical dimension. A ceremonial center of pyramid–temples (Fig. E-2) is only the final stage of its history (C). But what about its origins (A) and evolution (B)? In previous pages, we have seen that a surprising amount is known about the origin and evolution of such structures. Yet in much of the popular literature, what is deemed mysterious is only the final product (C). Rarely do we find in this literature the faintest attention given to where the idea for (C) began or how it came to be what it is.

We have talked about science and logic as a means of evaluating pseudo-scientific writings. In the case of archaeology, science can be applied to the material remnants of our past. Ruins can be probed, dated, and reconstructed, and thereby science provides explicit, although perhaps incomplete, descriptions of the past. But the real subject of archaeology is the men and women who have come before us. The patterns of the lives of people who created the great monuments of the past are generally not recoverable by archaeological science, nor are their motives and impulses. The forces responsible for shaping ancient cultures and inaugurating ponderous megalithic constructions may lie

Fig. E-1. Coming across this sculpture in a remote location, would future archaeologists correctly interpret its meaning and importance? *(Ashford Hollow Foundation for the Visual and Performing Arts)*.

more obscurely in ideology, philosophy, and religion. Science, in other words, should not be overrepresented as a means of deducing all our knowledge of man's ancient past. Nor should it be discarded as some sort of dinosaur incapable of dealing with so-called fresh, imaginative ideas and explanations for mysteries of the past or those surrounding us at present.

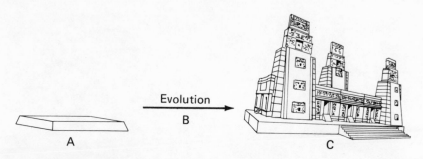

Fig. E-2. Structures such as (C) do not appear full-blown, but evolve from simpler forms such as (A) through several stages, subsumed under the heading "evolution" (B).

In conclusion, we hope we have carried the reader through an odyssey of mystery, both interesting and instructive. We lament the homage accorded to those mysteries so amenable to solution by logic and clear, critical thinking. Yet the day when all genuine mystery is dissipated, if such a day could ever come (and we doubt it), would be equally lamentable, for it has always been the attraction of mysteries, and the overwhelming urge to solve them, that have carried mankind from Stone Age campfires to exploration of the planets. In *The World as I See It,* Albert Einstein remarked (p. 5):

> The fairest thing we can experience is the mysterious. It is the fundamental emotion which stands at the cradle of true art and true science. He who knows it not and can no longer wonder, no longer feel amazement, is as good as dead, a snuffed-out candle.

# References

Atkinson, R. J. C. 1956. *Stonehenge*. Toronto: Collins.

Balsiger, D., and Sellier, C. 1976. *In Search of Noah's Ark*. Los Angeles: Sunn Classic Books.

Berlitz, C. 1969. *The Mystery of Atlantis*. New York: Nordon Publications.

Berlitz, C. 1972. *Mysteries from Forgotten Worlds*. New York: Dell Publishing Co.

*The Bible,* King James version.

Bord, J., and Bord, C. 1976. *The Secret Country*. New York: Walker and Co.

Borst, L. B., and Borst, B. M. 1975. *Megalithic Software*. Williamsville, N.Y.: Twin Bridge Press.

Bowman, J. S. 1971. *The Quest for Atlantis*. New York: Doubleday and Co.

Breasted, J. H. 1948. *A History of Egypt*. New York: Charles Scribner's Sons.

Brennan, J. H. 1975. *The Ultimate Elsewhere*. New York: New American Library.

*Buffalo Evening News,* August 1, 1966.

Camp, L. S. de. 1954. *Lost Continents*. New York: Ballantine Books.

Carter, G. F. 1975. *In Honor of Harold S. Gladwin*. Occasional Publications 2, no. 25. Arlington, Mass.: The Epigraphic Society.

Cathie, B. 1968. *Harmonic 33*. Wellington, New Zealand: A. H. & A. W. Reed.

Cathie, B., and Temm, P. 1971. *Harmonic 695*. Wellington, New Zealand: A. H. & A. W. Reed.

Cazeau, C. J., Hatcher, R. D., Jr., and Siemankowski, F. T. 1976. *Physical Geology*. New York: Harper and Row.

"Celtic Marks Found in New England." 1975. *The Courier Express* (Buffalo, N.Y.), August.

Charroux, R. 1967. *Masters of the World*. New York: Berkley Publishing Co.

Clark, J., and Coleman, L. 1975. *The Unidentified*. New York: Warner Books.

Cyr, D. L. 1977. "An Expedition to Painted Cave with Egerton Sykes." *Stonehenge Viewpoint* (Santa Barbara, Calif.) 8, no. 5.

Daniel, G. 1977. Review of *America B.C. The New York Times Book Review,* March 13, p. 12.

Dixon, R. 1928. *The Building of Cultures*. New York: Charles Scribner's Sons.

The editors of *Pensée*. 1976. *Velikovsky Reconsidered*. New York: Doubleday and Co.

Edwards, I. E. S. 1974. "The Collapse of the Meidum Pyramid." *Journal of Egyptian Archaeology* 60.

Einstein, A. S. 1949. *The World as I See It*. New York: New York Philosophical Library.

Evans, B. 1954. *The Spoor of Spooks and Other Nonsense*. New York: Alfred A. Knopf.

Ewing, M. 1949. "Discoveries on the Mid-Atlantic Ridge." *National Geographic* 96, no. 5, pp. 611–640.

Feldman, M. 1977. *Archaeology for Everyone*. New York: Quadrangle/The New York Times Book Co.

Fell, B. 1976. *America B.C.* New York: Quadrangle/The New York Times Book Co.

Fleming, T. 1977. "Who Really Discovered America?" *Reader's Digest,* February.

Fort, C. 1923. *New Lands.* New York: Ace Books.

Fort, C. 1931. *Lo!* New York: Ace Books.

Gaddis, V. 1965. *Invisible Horizons.* Philadelphia: Chilton Books.

Gill, G. 1976. "The Glendo Skeleton and its Meaning in Light of Post-Contact Racial Dynamics in the Great Plains." *Plains Anthropologist* 21, no. 72.

Gill, G. 1978. "Population Clines of the North American Sasquatch as Evidenced by Track Length and Estimated Stature." Paper delivered at the Conference on the Anthropology of the Unknown: Sasquatches and Similar Phenomena, Vancouver, May.

Godfrey, W. S. 1951. "The Archaeology of the Old Stone Mill in Newport, Rhode Island." *American Antiquity* 17, no. 2.

Goodavage, J. 1968. *Write Your Own Horoscope.* New York: New American Library.

Goodwin, W. B. 1946. *The Ruins of Great Ireland in New England.* Boston: Meador Publishing Co.

Green, G. 1967. *Let's Face the Facts about Flying Saucers.* New York: Popular Library.

Guenette, R., and Guenette, F. 1975. *Bigfoot: The Mysterious Monster.* Los Angeles: Sunn Classic Books.

Hammond, N. 1977. "The Earliest Maya." *Scientific American,* March.

Hawkins, G. 1965. *Stonehenge Decoded.* Garden City, N.Y.: Doubleday and Co.

Hole, F., and Heizer, R. F. 1973. *An Introduction to Prehistoric Archaeology.* New York: Holt, Rinehart and Winston.

Hoyle, F. 1972. *From Stonehenge to Modern Cosmology.* San Francisco: W. H. Freeman.

Hutin, S. 1970. *Alien Races and Fantastic Civilizations.* New York: Berkley Publishing Co.

Hynek, J. A. 1972. *The UFO Experience.* Chicago: Henry Regnery.

Ingstad, H. 1971. "Norse Sites at L'Anse aux Meadows." In *The Quest for America,* edited by G. Ashe. New York: Praeger.

James, T. G. H. 1972. *The Archaeology of Ancient Egypt.* New York: Henry Z. Walck.

Jeffery, E. C. 1952. *The Pyramids and the Patriarchs.* New York: Exposition Press.

Jeffrey, A.-K. T. 1973. *The Bermuda Triangle.* New York: Warner Books.

Keel, J. A. 1971. *Our Haunted Planet.* Greenwich, Conn.: Fawcett Publications.

Keyhoe, D. E. 1970. *Flying Saucers from Outer Space.* Universal-Tandem Publishing Co.

Keyhoe, D. E. 1973. *Aliens from Space.* New York: Doubleday and Co.

Kidder, A. V. 1960. "Wanted: More and Better Archaeologists." *Expedition 2,* no. 2.

Kolosimo, P. 1973. *Timeless Earth.* New York: Bantam Books.

Kusche, L. 1975. *The Bermuda Triangle Mystery—Solved.* New York: Warner Books.

Lucas, A., and Harris, J. R. 1962. *Ancient Egyptian Materials and Industries.* London: Edward Arnold.

McGervey, J. D. 1977. "A Statistical Test of Sun-Sign Astrology." *The Zetetic,* no. 2, pp. 49–54.

McIntyre, L. 1975. "Mystery of the Ancient Nazca Lines." *National Geographic,* May.

McKusick, M. 1970. "The Davenport Conspiracy." Report of the Office of the State Archaeologist, Iowa City.

Mendelssohn, K. 1971. "A Scientist Looks at the Pyramids." *American Scientist* 59, no. 2.

Mertz, H. 1972. *Gods from the Far East: How the Chinese Discovered America.* New York: Ballantine Books.

Métraux, A. 1971. Ethnology of Easter Island. Bishop Museum Reprints, *Bernice P. Bishop Museum Bulletin* 160 (original publication, 1940).

Mooney, R. 1974. *Colony: Earth.* Greenwich, Conn.: Fawcett Publications.

Mooney, R. 1975. *Gods of Air and Darkness.* Greenwich, Conn.: Fawcett Publications.

Mulloy, W. 1974. Contemplate the Navel of the World. *Americas Magazine* 26, no. 4.

"Mystery Hill Tour Guide Map." No date. North Salem, N.H.: Mystery Hill Corporation.

*New York Times,* July 3, 1954; June 5, 1965.

Nichols, E. 1975. *The Devil's Sea*. New York: Award Books.

"Objections to Astrology." 1975. *The Humanist* 35, no. 5.

Pauwels, L., and Bergier, J. 1960. *Morning of the Magicians*. New York: Avon Books.

Perry, W. J. 1923. *The Children of the Sun*. London: Methuen and Co.

*Playboy* Interview: Erich von Däniken and Timothy Ferris. 1974. *Playboy* 21, no. 8.

Pohl, F. J. 1952. *The Lost Discovery: Uncovering the Track of the Vikings in America*. New York: W. W. Norton.

Robinson, L. 1972. *Edgar Cayce's Story of the Origin and Destiny of Man*. New York: Berkley Publishing Corp.

Roggeveen, M. J. 1722. Extract from the official log of the voyage of Mynheer Jacob Roggeveen, in the ships Den Arend, Thienhoven and De Afrikaanische Galey, in 1721–22, in so far as it relates to the discovery of Easter Island. *Haklyut Soc.*, 2nd ser., no. 13, Cambridge, 1908.

Routledge, S. 1919. *The Mystery of Easter Island*. London: Sifton, Pread.

Sagan, C., and Page, T., eds. 1972. *UFO's—A Scientific Debate,* New York: W. W. Norton.

Schul, B., and Pettit, E. 1975. *The Secret Power of Pyramids*. Greenwich, Conn.: Fawcett Publications.

Skelton, R. A., Marston, T. E., and Painter, G. D. 1965. *The Vinland Map and the Tartar Relation*. New Haven and London: Yale University Press.

Smith, G. E. 1923. *The Migrations of Early Culture*. Manchester, England: Manchester University Press.

Smith, W., ed. 1974. *Predictions for 1975*. New York: Award Books.

Smith, W. 1975. *The Secret Forces of the Pyramids*. New York: Kensington Publishing Corp.

Spencer, J. W. 1969. *Limbo of the Lost*. New York: Bantam Books.

Spencer, J. W. 1974. *No Earthly Explanation*. New York: Bantam Books.

Spencer, J. W. 1975. *Limbo of the Lost—Today*. New York: Bantam Books.

Steiger, B. 1974. *Mysteries of Time and Space*. New York: Dell Publishing Co.

Steiger, B., ed. 1976. *Project Blue Book*. New York: Ballantine Books.

Story, R. 1976. *The Space-Gods Revealed*. New York: Harper and Row.

Thomas, L. 1930. *India: Land of the Black Pagoda*. New York: P. F. Collier & Son Corp.

Tomas, A. 1971. *We Are Not the First*. New York: Bantam Books.

Toth, M., and Nielsen, G. 1974. *Pyramid Power*. New York: Freeway Press.

Trefil, J. 1978. "A Consumer's Guide to Pseudoscience." *Saturday Review,* April.

Trench, B. Le P. 1960. *The Sky People*. New York: Award Books.

Umland, E., and Umland, C. 1974. *Mystery of the Ancients*. New York: New American Library.

Vail, I. N. 1977. *Canopy Skies of Ancient Man*. Cited in *Stonehenge Viewpoint* (Santa Barbara, Calif.) 8, no. 5.

van der Veer, M. H. J. Th., and Moerman, P. 1972. *Hidden Worlds*. New York: Bantam Books.

van Sertima, I. 1976. *They Came before Columbus*. New York: Random House.

Velikovsky, I. 1950. *Worlds in Collision*. New York: Doubleday and Co.

Velikovsky, I. 1952. *Ages in Chaos*. New York: Doubleday and Co.

Velikovsky, I. 1955. *Earth in Upheaval*. New York: Doubleday and Co.

Velikovsky, I. 1977. *Peoples of the Sea*. New York: Doubleday and Co.

Vescelius, G. S. 1955. "North Salem, N.H., Site Excavations, Report to Early Sites Foundation." Typewritten manuscript.

von Däniken, E. 1969. *Chariots of the Gods?* New York: Bantam Books.

von Däniken, E. 1972. *Gods from Outer Space*. New York: Bantam Books.

von Däniken, E. 1972. *The Gold of the Gods*. New York: Bantam Books.

von Däniken, E. 1974. *Miracles of the Gods*. New York: Dell Publishing Co.

Wallace, B. L. 1971. "Some Points of Controversy." In *The Quest for America,* edited by G. Ashe. New York: Praeger.

"Washington Whispers" (column). 1977. *U.S. News and World Report* 84, no. 16.

Watson, L. 1973. *Supernature*. New York: Doubleday and Co.

Wauchope, 1962. *Lost Tribes and Sunken Continents*. Chicago: University of Chicago Press.

Wernick, R. 1973. *The Monument Builders*. New York: Time-Life Books.

White, P. 1974. *The Past Is Human*. New York: Taplinger Publishing Co.

Whorf, B. L. 1940. "Decipherment of Maya Hieroglyphs." In *Language, Thought, and Reality: Selected Writings of Benjamin Lee Whorf,* edited by J. B. Carroll. Cambridge, Mass.: M.I.T. Press.

Wilson, C. 1972. *Crash Go the Chariots*. New York: Lancer Books.

Wilson, C. 1974. *UFOs and Their Mission Impossible*. New York: New American Library.

Winer, R. 1974. *The Devil's Triangle*. New York: Bantam Books.

Wolff, W. 1948. *Island of Death. A New Key to Easter Island's Culture through an Ethnopsychological Study*. New York: J. J. Augustin.

Zinn, J. 1973. "The Mystery of Mystery Hill." *New Hampshire Echoes* (Concord, N.H.), Sept.– Oct.

# Index

The numbers of pages on which illustrations appear are in *italics*.